A Profile of the Steel Industry

A Profile of the Steel Industry

Global Reinvention for a New Economy

Peter Warrian

A Profile of the Steel Industry: Global Reinvention for a New Economy

First published in 2012 by
Business Expert Press, LLC
222 East 46th Street, New York, NY 10017
www.businessexpertpress.com

ISBN-13: 978-1-60649-417-2 (paperback)

ISBN-13: 978-1-60649-418-9 (e-book)

DOI 10.4128/9781606494189

Business Expert Press Industry Profiles collection

Cover design by Jonathan Pennell
Interior design by Exeter Premedia Services Private Ltd.,
Chennai, India

First edition: 2012

10 9 8 7 6 5 4 3 2 1

Printed in the United States of America.

Abstract

The steel mill and the auto plant were the icons of the 20th-century industrial economy. By the early 21st century, many people viewed steel as Big, Ugly, and Gone to China. Not so. In manufacturing, The World Is Not Flat! The steel industry has continuously been forced to remake itself. This book describes those developments and dynamics.

Keywords

steel, manufacturing, technology, trade, unions

Contents

Foreword

Globalization and Technical Change

We live in a time of change. In particular, we live in a world in which computing power is steadily transforming the way people work. Today, there are many who have never seen anyone operate an elevator as a way to earn a living because that task was long since delegated to a button, a sensor, and a computer chip. If we were to track occupations as we do animal species, secretaries would surely be joining elevator operators and many others on the ever-lengthening list of work species endangered or displaced by the impact of technology.

Improvement in the cheapness and ease of sending and receiving information has also changed how corporations work. This has made tight coordination with far-flung global enterprises much more feasible. Container ships, huge freighters, and air freight have also made moving not only information but also goods between national geographies cheaper and easier. These changes, by opening up national markets to international trade, have worsened the fate of industries and the divisive public policy issue, rather than something that is in the main naturally arranged by geographical separation. Successful remote market access, with all its possibilities for disruptive change, now occurs on a much larger scale than ever before.

These changes are usually described as globalization. But globalization is a very abstract word; it is not clear that it means the same thing to different people. Nor can it, because, in fact, globalization is very different in its effect on different industries. Simple models of what is going on cannot be expected to be realistic.

What then can be done to make these important changes and their consequences more intelligible? I suggest that one thing to do is to obtain a realistic view of at least one industry. I do not mean that one sample industry can tell you what globalization means to other industries. Rather when you have seen what can happen in one industry, and have seen that much that actually happened is neither intuitive nor obvious, you will be in a better position to ask and comprehend more about what is going on in other industries. This will help in understanding even if the actual events occurring in the other industry are quite different.

In some manufacturing industries, access to necessary natural inputs may be easy, and an insignificant advantage or disadvantage to firms in different locations, but in others it can be a life or death issue. Steel, surprisingly enough, has experienced both these situations.

This book describes the steel industry; it gives its history; it describes the many ways that steel is made today, how it was made in the past, and how it may be made in the future. It describes the effect of change. We watch as the Bessemer process transforms steel from something rare, special, and precious, like the "Damascus steel" prized for sharp-edged swords, into something common: an everyday material with a thousand uses. And that first seminal event turns out to be only a beginning.

We watch as steel plants turn from being assemblages of burly men sweating against a background of flames and hot glowing steel to vast buildings containing only a few people. And these few are perched high above the hot plant floors, looking down at what is happening below, and looking toward computer screens. From their lofty perches they operate the controls that move the hot steel shapes below from place to place and are in control of every detail of the processing.

We watch as almost all the corporations that would be even slightly familiar to Americans, corporations like Bethlehem Steel, Jones and Laughlin, Lackawanna Steel, Inland Steel, and Republic Steel, disappear through mergers and end up being parts of something called ArcelorMittal. ArcelorMittal is not an American household word, but today it is the world's largest single steel producer, a widespread global corporation controlled by two members of India's wealthy Mittal family.

Meanwhile, the company that was once a household word, U.S. Steel, and that was also once the largest company in the world, still survives after a merger with Carnegie Steel. But it has been reduced in scope and scale with a production capability less than a quarter of that of ArcelorMittal.

How did these enormous changes come about? There are many reasons, but the most important are certainly not what a purely economic overview would predict. For example, in steel, where labor is a small part of cost, cheap labor was not the issue. Much of the difficulties of American companies in the 1980s and 1990s were not the simple working out of something economically inevitable.

Technical decisions played a major role. And those decisions were shaped in part not only by financial pressures but also by personal histories and national differences. Innovative Japanese firms, supported by the Japanese government's Ministry of International Trade and Industry (MITI), pioneered in better steel manufacturing processes, but most US firms, sticking with the familiar, fell behind. Organizational issues too played a role.

Then there was and is the rise of China. While many US integrated steel firms were acting on the appealing idea of concentrating on their core competencies and were busy selling off their "peripheral" coal mines and iron ore sources, China was busy acquiring them. Then with the development of both the appetite for steel and the ability to make steel—China today makes more than half the world's steel—raw material prices went through the roof handicapping those who did not have raw materials in house.

In addition, there is the rise of the minimills that made steel from scrap, and the vital role played throughout by union-management relations. This book describes in some detail what was at stake in the great steel strike of 1959 and the later landmark steel agreements, and what their consequences actually were.

This book contains all this and much more. Its account of steel industry shows clearly how much more there is to a real industry than is apparent from a distance; and how that much more can be decisive.

We are fortunate that the steel industry is described to us here by Peter Warrian. Peter Warrian, a Canadian, has been in the steel industry for 35 years. He is a steel insider with deep knowledge, but he is not a simple partisan of management, or of labor, or of government. He has been the chief economic advisor to the Ontario government, has held important posts in the steelworkers union, and has consulted with major steel corporations. He is now a research fellow at the University of Toronto and head of the Lupina Foundation.

Peter Warrian's account of the steel industry does not make the events he describes seem mysterious, or due to greed, or to stupidity. Rather he describes humans working as best they know how, but embedded in an environment that is changing, complex, and unpredictable.

That is the kind of account we need if we are going to understand and be realistic about the role of industries in a globalized world.

Ralph Gomory

Abbreviations

AERA	American Railway Engineers Association
ASTM	American Society for Testing Materials
BOF	Basic Oxygen Furnace
BRIC	Brazil, Russia, India, and China
BWE	Bull Whip Effect
CCs	Continuous Casters
CCSC	Coordinated Committee Steel Companies
COLA	Cost-of-Living
CPA	Cooperative Partnership Agreements
CWS	Co-Operative Wage Study
DRI	Directly Reduce Iron
EAFs	Electric Arc Furnaces
ENA	Experimental Negotiating Agreement
ERISA	Employee Retirement Income Security Act
GATT	General Agreement on Tariffs and Trade
GPN	Global Production Network
ICT	Information and Communications Technology
ISO	International Standards Organization
ITC	International Trade Commission
KIBS	Knowledge Intensive Business Services
LMPT	Labor Management Participation Teams
MITI	The Ministry of International Trade and Industry
OBC	Oxygen Blown Converter
OEMs	Original Equipment Manufacturers
OH	Open Hearth Furnace
PBGC	Pension Benefits Guarantee Corporation
PBGF	Pension Benefits Guarantee Fund
SAE	Society of Automotive Engineers
SCC	Sustained Contingent Collaborative
SPC	Statistical Process Control

TRIP	Transformation-induced plasticity steel
TWIP	Twinning-induced plasticity steel
UPMC	University of Pittsburgh Medical Center
VRAs	Voluntary Restraint Agreements
WTO	World Trade Organization

CHAPTER 1

Introduction

The steel mill and the auto plant were the icons of the 20th-century industrial economy. By the 21st century, many people viewed steel as Big, Ugly and Gone to China. An informed observer would have the picture that in the 1980s the traditional big steel mills were overtaken by the modern smaller "minimills." Then, after 2000 the whole business shifted to China. Not so. In manufacturing, The World Is Not Flat! Public discourse about globalization is dominated by simple models such as the existence of Chinese low wages means all manufacturing will inevitably re-locate there. The real world of industry is much more complicated and even hopeful. The steel industry has continuously been forced to remake itself, which it has done. This book describes those developments and dynamics.

Economic discourse in our time is dominated if not preoccupied with the issue of innovation. This is seen rightly to be the key to 21st-century competitiveness. The somewhat counter-intuitive facts are that, notwithstanding the stunning accomplishments of modern biomedical technologies, over 80% of innovation in the economy takes place in manufacturing.[1] Steel is the material backbone of manufacturing. So, in turn, the underpinning of manufacturing innovation lies in materials science and engineering, whose base is in steel.

The rise of China was "the" steel story of the past decade. It now has about 45% of the world's capacity, and taking the Recession into account, China may have close to half the operating capacity currently producing. But the rise of China has not fundamentally changed how steel is produced. What it has done is dramatically change the input prices for producing steel, primarily for iron ore, coal, and scrap. This has had a destabilizing effect on how the industry operates comparable to the disruptive Steel Trade Wars of the 1980s and 1990s.

There has also been a more subtle but critical technical change. With the rise of China, integrated steelmaking has risen again to the forefront of innovation in steel. Instead of further eating into integrated market share, minimills are now faced with equally challenging cost pressures from the dramatic rise in scrap prices as well as escalating energy prices. Energy prices alone now equal or exceed labor costs in steel production. And, at the bottom end of the commodity food chain, new micro-mills are starting to do to minimills what minimills did to integrated mills 40 years ago.

Internationally, the Brazil story may overtake the China story in steel in the next decade. Brazil's near monopoly on the highest grades of iron ore in addition to its policy agenda of controlling more of the value chain means that the boundary between steelmaking and raw materials may be redrawn at the slab stage. This could have a potentially greater impact globally on how steel is made and marketed than the rise of China.

Who Needs Steel?

We all need steel. It is the backbone material for manufacturing. We still live in a world where we primarily consume "stuff." It is estimated that 80% of consumer goods contain steel. In addition, if you are concerned about the environment then there should be more steel in your life. It is the most recycled material in the economy; 70% of steel is recycled compared to about 12% of aluminum. Steel's physical properties can withstand almost endless recycling.

Steel companies and national industries have also had to reconstruct themselves due to changing political, economic, and technology factors. For instance, the European Union was founded on the European Coal and Steel Community. Steel had been the cornerstone of modern warfare and Europe sought to put its troubled past behind it by seeking to unify this industry on a continental basis as a means to peace and prosperity.

There have been three major destabilizing forces in the past 50 years that have challenged steel companies and steel management around the world.

The first major postwar steel destabilization factor was trade. For the first half of the 20th century, steel was a stay at home industry. However, every emerging country wanted its own flag, airline, and steel mill. The rebuilt Japanese steel industry was the first that had a capacity that well exceeded domestic demand and required large volumes of exports in order to be profitable. This was the fundamental, though not the only, backdrop to the Steel Trade Wars of the 1970s and 1980s.

The second disruptive change was in technical steelmaking itself and the rise of the so-called minimills. These were small, more efficient mills making steel totally from scrap. The U.S. steel industry became ground zero for commercialization of this new steelmaking technology. They started at the commodity end of the market, producing rod for the construction industry. At first this was a minor irritant for big existing steel companies. But, by the later 1980s technology development by the minimills allowed them to move into the higher value added and higher margin flat-rolled steel product markets such as automotive. Just prior to the 2008 Financial Crisis, minimills were producing half the steel in the United States.

The third impact involves the rise of China. Virtually no one in the traditional steel industry saw it coming. China, with a stunning level of growth in the past decade, now has the ability to determine the marginal price of flat-rolled products around the world and has emerged as a significant player in export markets. However, the real disruptive impact of China has been on raw material prices: iron ore and coal.

Why People Should Read This Book

People should read this book because steel continues to be a huge factor in the economy.

The American steel industry directly employed more than 139,000 workers and contributed $17.5 billion in value added to gross domestic product in 2011. In addition, it contributed to household incomes by generating $121.4 billion in wages and salaries across the economy, the purchase of $20 billion in materials from other industries, $8 billion in services, $5 billion in energy products, $4.5 billion in machinery, $4.4 billion in wholesale and retail trade.

The total economic impact of steel in the economy is even larger. Its net contribution to the overall economy was $246 billion. Every job in steel creates seven jobs in the US economy. In total, the steel industry supports 1,022,009 jobs across the economy as of 2011.[2]

What kinds of other industries and jobs does the steel industry support? Economists use demand multipliers to estimate how much a change in one industry has on another industry, for example, how much an increase in the level of activity in the steel industry will have on demand in the machinery industry. Readers would not be surprised to learn that steel has important impacts on industries such as machinery and mining. They would be surprised to learn that steel has an even greater impact on jobs in professional, scientific and technical services, education and health care.

Total economic activity supported by steel, including things such as the impact of taxes paid by the steel industry, produces an 11.298 employment multiplier. For all these reasons, people who care about the economy and its future should try to understand the Steel Story.

There are two individual profiles for the kind of readers who would benefit from reading this book. First is a student in a managerial economics class to whom it might be assigned as a supplemental reading. The second is a person who is an investor who knows little about it and reads the book to get a better grasp on the steel industry. However, there is a much broader audience of people concerned about the implications of globalization of the economy. For them, learning about this specific industry may suggest lessons for a perspective on the future of many industries.

For instance, profit is critical to steel companies. Without it they would not be viable. At the same time, the challenges and decisions confronting decisionmakers in the industry are not driven by short-termism and simplistic "shareholder value" ideologies. Companies are facing on-going struggles to continue to exist, keeping up with technological and market changes, making better steel, improving quality and productivity. Even in the heyday of Big Steel, profit was a central but not exclusive issue for steel executives, albeit that they were riding a wave of postwar expansion, oligopolistic pricing and an absence of significant competition. All of this changed in the 1970s. New competitors, loss of price leadership, new regulatory challenges from the environmental and employee benefits side all hit the large integrated producers at once. The 1990s saw a selling off of assets and de-verticalization in the name of realizing short-term shareholder value. The rise of China and the impact on raw materials destabilized the reconstituted mills. The new steel ownership around the world understands that it will realize shareholder value only if it gets the business operations, technology, marketing and human resources right. This is a particular challenge for large, capital intensive, environmentally exposed companies like steel. However the lessons extend to all other industries in the new global economy.

The Perspective of This Book

It is the perspective of this book that the global steel industry in the past decade is undergoing its second great re-organization in the past 75 years.

The first was the post-World War II steel restructuring under the direction of the Allied Powers as described in Gary Herrigel, *Manufacturing Possibilities* (2010). He challenges the traditional view of the steel industry in the postwar period. The pre-World War II steel industries of Germany, the USA, and Japan were all characterized by a dominant integrated producer surrounded by smaller, specialized firms. Following the war, they were decentralized into an industry dominated by a small set of mass production, oligopolistic, large steel companies. But in the 1970s and 1980s, in all leading countries, traditional integrated operations have been rationalized and minimills deployed a radically disruptive steelmaking technology to gain a major market share. By the first decade of the 21st century, the postwar steel industry stood on its head. A limited number of highly specialized integrated firms were surrounded by standardized minimill producers.

The subject of this book is the next stage of these developments. The global restructuring of steel in the past decade has produced a new configuration of core specialized firms and secondary commodity firms. How are these core primary steel producers now being managed and coordinated after the global steel consolidation? And, how do steel companies learn? National steel industries have become merged into global supply chains. One's position in a contingent collaborative supply chain is correlated with learning capabilities. New global steel companies are clearly embedded within this framework.

Globalization and the Steel Industry

We will be discussing steel as a globalized industry.

Three phenomena are at the heart of globalization. The first is the internationalization of ownership structures. This is most evident over the past several years, among steel-producing companies where there has been a change of ownership in virtually every steel-producing operation. This change in ownership in most cases involved a shift from domestic control to international control.

A direct consequence of the transformation of ownership structures is significantly increased international flows of not only capital and technology, but also managerial norms and talent. While attention most often focuses on capital and technology, the adoption of international managerial norms and the international flows of talent may have the most far-reaching impact on operations management. These include productivity benchmarking, new approaches to work organization and distinct strategies related to training and human resources development.

The second phenomenon that defines globalization is the accelerated integration of international markets on a regional basis such as the North American Free Trade Agreement (NAFTA) steel market. The signature development that marks this integration is a sharp increase in *both* exports and imports of steel products. This process accelerated in the past decade and is the basis for many people's perception that the industry has "disappeared." In reality, one kind of industry disappeared. But it was replaced by another.

The merger wave in the industry resulted in major changes to facilities: some were closed, others expanded, many became more specialized. This is reflected in the increased export orientation of the industry. The rationalization of capacity drove a step-function increase in productivity that reduced overall employment and is also reshaping the skill needs of the steelworker of the future, as discussed in Chapter 11. Although globalization led to an increase in the steel industry's export orientation, it would be a serious error to discount the continuing importance of the domestic market. Even with exports taking an increased share of output, the domestic market still accounts for at least half of industry shipments. As discussed in Chapter 7, the industry is still vulnerable to unfair trade practices, such as "dumping."

The third phenomenon at the heart of globalization is a marked increase in the scope and reach of international supply chains. The logic driving the globalization of supply chains is the competitive advantage that derives from achieving an approximate symmetry between labor intensiveness and skill availability on the one hand and labor cost on the other. The migration of low-skill, labor-intensive production processes to lower cost jurisdictions is not a new story. What is new is the increase in the scope and reach of international supply chains to include production processes that were much less vulnerable to offshoring prior to this decade.

In summary, for steel production, the impact of globalization is most evident in the internationalization of ownership, the rationalization of production and the adoption of international performance benchmarking. These developments have fundamentally altered productivity conditions over the past five years and will reshape human resources requirements in the industry in the future. Labor costs have never been a static figure in the industry, they vary over time. Currently, owing to the relatively low share of labor in total production costs (approximately 15%), the industry is much less vulnerable to pressure to relocate production to low-cost jurisdictions. However, the internationalization of production has increased the intensity of competition for investment and technology within global companies. This puts new levels of pressure on workplace performance.

How This Book Is Organized

The book tells the story of a huge, complicated industry. It begins with an introduction to the making of steel, a brief history of the metal and its usage and then a quick summary of the history of the industry in the past century. As we proceed through the topical chapters, there is inevitably movement back and forth between historical periods and developments of companies and technologies. Unfortunately, it is not possible to do all of the history in one place. Often, a chapter will begin with a description of a current situation or fact that will strike the reader as unfamiliar or counter-intuitive. The chapter will then go back in time to give the history behind the facts to make the current situation more understandable for the reader. Utmost effort is made to make the context clear whenever there are such references.

Chapter 2 seeks to give the general reader an introduction to Steel Basics, the metal, its production and steel companies in a historical context.

Chapter 3 describes how steel companies operate, the objectives of executives, what global consolidation has meant in terms of benchmarking and coordination. Because steel is an intermediate product, it also describes steel supply chains, upstream to raw materials, and downstream to steel-consuming industries.

Chapter 4 talks about how steel industries operate and how they have been re-organized over time. It also describes how steel firms are positioned within regional markets and in the global economy.

Chapter 5 describes how steel product pricing has evolved and the new sources of competition, pressures from imports, and by the minimills. It also discusses the impact of surging raw material prices, principally coal and iron ore, on the costs and margins of steel companies.

Chapter 6 discusses the challenging history of steel industry labor relations, which have shown steel unions and management on their best and their worst days. It also identifies human resource issues including legacy health and pension costs, and the need to find new forms of employee engagement and leveraging of production knowledge on the shop floor.

Chapter 7 summarizes the troubling story of steel trade disputes, particularly the vexatious issues of foreign dumping and antidumping trade policies as applied to steel. It also discusses the rise of China as a new challenge for global steel and, as illustrated by the Chinese case, the continued role of the state in the steel industry.

Chapter 8 examines steel and external market forces. Is there a steel growth story? How does the industrial cluster associated with steel facilities work? It concludes with a discussion of new global steel and manufacturing supply chains.

Chapter 9 discusses steel technology developments and its evolution from company R&D facilities to traded knowledge to global consortia. It also situates steel in the future of materials competition.

Chapter 10 discusses regulation of the steel industry from antitrust policy to environmental policy. It also describes the importance of industrial standards such as SAE and ISO.

Chapter 11 discusses steel in the postindustrial, knowledge-based economy. Specifically, it looks at how steel companies learn and how knowledge networks arise between steel companies and their customers.

Chapter 12 concludes with some general observations about how further restructuring of the steel industry may unfold, its linkages to the future of manufacturing, materials competition in the new economy and the potential impact of public policy on the industry.

CHAPTER 2

Steel Basics

Steel: The Metal and Its History

Iron is found in the Earth's crust in the form of an ore, usually an iron oxide. Iron is extracted by smelting, removing the oxygen, and combining the ore with a preferred chemical partner such as carbon. Ancient methods of smelting have been used since the Bronze Age. The Haya people of East Africa invented a type of high-heat blast furnace, which allowed them to forge carbon steel nearly 2,000 years ago.

Steel is a metal alloy made by combining iron and another element, usually carbon, but it can also be combined with manganese, chromium, vanadium, and tungsten. When carbon is used, its content in the steel is between 0.2% and 2.1% by weight, depending on the grade. Varying the amount of alloys and their form controls qualities such as the hardness of the final product. Steel with increased carbon content can be made harder and stronger than iron, but such steel is also less *ductile* than iron, meaning that it is less able to be stretched into something like wire. Alloys with carbon content higher than 2.1% are known as cast iron.

The Earliest Sites of Steelmaking Were in the Middle East, East Africa, and Far East

The earliest known production of steel is a piece of ironware excavated from an archeological site in Anatolia and is about 4,000 years old. Other ancient steel comes from East Africa, dating back to 1400 BC. In the 4th century BC, steel weapons were being used by the Roman military. The Chinese of the Warring States (403–221 BC) had quench-hardened steel, while the Chinese of the Han Dynasty (202 BC–220 AD) created steel by melting together wrought iron with cast iron.

Evidence of the earliest production of high carbon steel on the Indian Subcontinent was found in Sri Lanka. Wootz steel was produced in India by about 300 BC. In Sri Lanka, this early steelmaking method employed the unique use of a wind furnace, blown by the monsoon winds capable of producing high-carbon steel. Also known as Damascus steel, Wootz is famous for its durability and ability to hold an edge. It was essentially a complicated alloy with iron as its main component. Along with their original methods of forging steel, the Chinese also adopted the production methods of creating Wootz steel, an idea imported into China from India by the 5th century AD.

In Europe, steel was produced long before the Renaissance, but it became more common after the 17th century. With the invention of the Bessemer process in the mid-19th century, steel became an inexpensive mass-produced material. Further refinements in the process, such as the basic oxygen furnace (BOF), lowered the cost of production while increasing the quality of the metal. Modern steel is generally identified by various grades defined by assorted standards organizations such as the Society of Automotive Engineers (SAE).

The modern era in steelmaking began in England with the introduction of Henry Bessemer's process in 1858. His raw material was pig iron. This enabled steel to be produced in large quantities cheaply. The Gilchrist–Thomas process (or *basic Bessemer process*) was an improvement to the Bessemer process, lining the converter with a basic material to remove phosphorus.

These were rendered obsolete by the Linz–Donawitz process of basic oxygen steelmaking (BOF), developed in the 1950s. Basic oxygen steelmaking is superior to previous steelmaking methods because the oxygen pumped into the furnace limits impurities. Now, electric arc furnaces (EAFs) are a common method of reprocessing scrap metal to create new steel. They can also be used for converting pig iron into steel, but they use a lot of electricity (about 440 kWh per metric ton), and are thus generally only economical when there is a plentiful local supply of cheap electricity.

Modern steels are made with varying combinations of alloy metals. Carbon steel, composed simply of iron and carbon, accounts for 90% of steel production. High-strength, low alloy steel has small additions (usually <2% by weight) of other elements, typically 1.5% manganese, to provide additional strength for a modest price increase. Low alloy steel is combined with other elements, usually molybdenum, manganese, chromium, or nickel, in amounts of up to 10% by weight to improve the hardenability of products like beams. Stainless steels and surgical stainless steels contain a minimum of 11% chromium, often combined with nickel, to resist corrosion and rust. Some stainless steels are magnetic, while others are not.

Other more modern steels include tool steels, which are alloyed with large amounts of tungsten and cobalt or other elements to maximize hardening. This also allows the use of precipitation hardening and improves the alloy's temperature resistance. Tool steel is generally used in axes, drills, and other devices that need a sharp, long-lasting cutting edge. Other special-purpose alloys include weathering steels, which weather by acquiring a stable, rusted surface, and so can be used unpainted.

Many other high-strength alloys exist, such as dual-phase steel, which is heat-treated to contain a microstructure that gives extra strength. This bridges the age-old problem where increased strength means increased hardness, which is what brought down the Titanic. Particularly in cold water, the ship's steel was unusually brittle and could not absorb the impact of contact with ice.

Transformation-induced plasticity (TRIP) steel involves special alloying and heat treatments to create a very strong but still malleable metal. Twinning-induced plasticity (TWIP) steel uses a specific type of strain to increase the effectiveness of work hardening on the alloy. Others use a combination of over a dozen different elements in varying amounts to create a relatively low-cost metal for use in high-wear applications such as bulldozer blade edges and cutting blades on the Jaws of Life.

Before the introduction of the Bessemer process and other modern production techniques, steel was expensive and was only used where no cheaper alternative existed, particularly for the cutting edge of knives, razors, swords, and other items where a hard, sharp edge was needed. It was also used for springs, including those used in clocks and watches. With the advent of speedier and lower cost production methods, steel has been easier to obtain and is much cheaper. It has replaced wrought iron for many purposes.

Images and Realities of the Steel Industry

Our perceptions of the steel industry and its presentation in the media are greatly determined by images, often those of the past. Therefore, it is important to get a quick read on the stages the industry has gone through.

The images of modern steel are still very much the creature of the early 20th century. This is the industry as it was portrayed in the classic film clips of sparks, smoke, and gritty workers appearing in grainy old movies. Sweaty men move around between molten metal and huge machines. It was a kind of industrial gigantism. Huge blast furnaces made hot iron, which was converted to steel in open hearth (OH) furnaces then rolled into flat and round shapes. In old pictures, you can see huge mechanical cranes and rolling machines moving hot metal around, though this actually represented the change from steam to electric power. The electrical machinery could be controlled much more finely and continuously than the older steam-driven processes. The biggest product lines were rails for transportation and building beams for the new generation of skyscrapers, which were a steel invention. In fact, the design of skyscrapers first appearing in Chicago was an iron railway bridge turned on end.

New Deal Steel

The interwar period was characterized by a change in product lines to more consumer goods—cars, refrigerators, and stoves. These required large scale production of sheet steel. The critical technology breakthrough was the continuous wide strip mill, which could continually pass metal back and forth under the rollers to produce the required gauges. No more did steelworkers have to grab hot sheets with tongs and passed them back and forth, one by one, into the rolling machines. This was the enabling technology for mass market steel.

Postwar Steel

The steel industry grew enormously during World War II to support the war effort. Technically, it was more of the same but focused on heavy plate products for ships, tanks, and guns. By the 1950s there was a fundamental shift to ramp-up the scale and speed of production to meet the requirements of the exploding consumer economy. New Basic Oxygen Furnaces (BOFs) at the steelmaking stage and continuous casters (CCs) revolutionized the rolling process. Together, this technology combination, only really perfected by the Japanese in the mid-1960s, became the defining technology for large steel producers. However, the EAF, a small scale but highly efficient scrap steel-fed technology began, in the 1960s, to eat into the lower value-added product markets.

Global Steel

Steel went global beginning in the 1990s. A wide consolidation of steel companies was sparked by the divestment of formerly government-owned steel mills to private ownership. Some were shut, but many were consolidated into new steel conglomerates, led by what is now the largest steel company in the world, ArcelorMittal. This was accompanied by the almost entirely unexpected development of China as the dominant steel industry, which in 2012 is expected to have nearly 900 MTs of steel capacity in a global steel industry of 1,600 MTs.

How Steel Is Made

There are two fundamentally different ways to make steel—directly from raw materials or by re-melting steel scrap. However, both are fundamentally batch processes, which, as discussed later, present challenges in a world of Just-in-Time delivery and lean production manufacturing.

Integrated steel companies start with blast furnaces making hot iron from iron ore and coal. The iron is then converted into steel in a basic oxygen furnace (BOF) also called oxygen blown converter (OBC) technology. These furnaces convert raw materials (iron ore, coking coal, and limestone) into basic steel shapes.

The second steelmaking technology uses electric arc furnaces (EAFs) with large electrodes which melt scrap to produce steel, essentially a recycling process (Figure 2.1). The scrap metal is then converted to basic steel shapes.

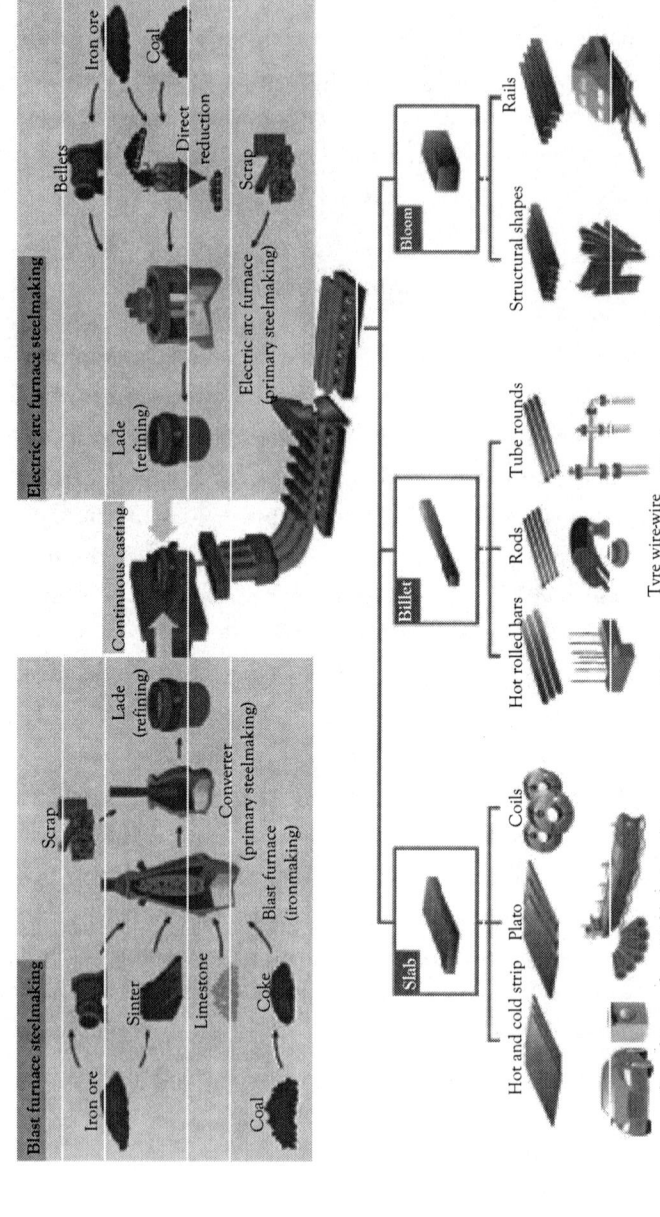

Figure 2.1. Steel manufacturing.

Source: World Steel Association.

There are some sites that use both technologies within the same facility. Today, in the United States, 48% of the EAF capacity is actually owned by integrated steel companies.

Primary steel mills produce the basic steel shapes: slabs, billets, and blooms, which are then further processed into semifinished steel products. Slabs are converted into hot and cold strip steel, steel plate, and coiled steel sheet. Billets are further processed into various shapes of bars, rods, and tubes. Blooms are made into structural shapes for the construction industry and for rail. The conversion of basic steel shapes into semiprocessed steel products usually takes place within a primary steel mill.

The following graphic illustrates the manufacture of the three basic steel shapes and their conversion into semiprocessed products for applications in various downstream industries.

Semiprocessed steel products may be further treated to produce products such as steel pipe. However, most semiprocessed steel products are supplied to downstream manufacturers. Historically, some of these downstream manufacturers were closely linked to primary steel producers, although this is no longer typically the case.

The downstream market may be supplied either directly or indirectly. Large industries, such as the auto industry, are supplied directly by the primary steel producers. Re-bar for use in reinforcing concrete is also usually supplied directly to the construction industry by the primary producers. In North America, companies in other industries now are generally supplied through Metal Service Centers, which are independent wholesale distributors of semiprocessed steel products (and other metals). Outside of North America, metal service centers are often affiliated to primary steel-producing companies and distribute mainly or exclusively the output of their parent firm.

These segments taken together, the primary steel mills of both BOF and EAF production, the service centers and steel fabricators, constitute the Steel Sector of the modern economy.

CHAPTER 3

How Steel Companies Operate

The primary audience for this book is students in MBA and Executive Education programmes. For this reason, much of the book is directed at issues of management, both within steel companies and more broadly. How do steel companies operate, what motivates executives in charge of them, and how do they manage and coordinate their huge and complicated global operations?

Classic steel companies were operated like huge centrally controlled machines. They had their own iron ore and coal mines, railroads, furnaces, rolling mills, and distribution centers. With this scale and throughput they produced steel products, which in their standard grades, cost less than the price of chicken at the local grocery store. Price pressures from low-cost minimills and foreign imports, along with financial pressure to maximize shareholder value, caused the integrated steel companies to change how they operated in the 1980s and 1990s. They closed or sold off most "noncore" assets such as mines, transportation, and distribution facilities. Instead of internally regulated functions, these became arms-length contractual relationships. In retrospect, the timing could not have been worse. The steel companies had become extremely vulnerable to escalating input cost pressures as the price of raw materials surged, driven by the emergence of Chinese steel demand. It caused a quadrupling and more in commodity prices, plus a level of volatility that was unprecedented that destabilized existing steel company operations. Managing dispersed supply chains instead of in-house operations became the new norm for steel companies. They joined other manufacturers in coordinating a level of complexity that was a qualitative change from the industrial organizations and culture of previous company operations. Steel companies are now immersed in the global system of networked manufacturing. This raises the question of what the steel company of the future will look like. Some commentators suggest that they will take on the role of systems integrators as have other advanced manufacturing companies. Maybe, maybe not.

The chapter begins with the current situation, who the new steel companies are, and brief profiles of the major players. It then turns to the history and culture of steel companies in the past century in order to outline the challenges older corporate structures and cultures present for the new generation of global steel management. We then return to the present and look at the objectives of steel company executives and how they measure results through benchmarking of local operations. Finally, we look at steel company operations as coordinated global supply chains, taking a closer look at the example of metallurgical coal. The chapter concludes with some suggestions of what the steel company of the future might look like.

Global Consolidation of Steel Companies

We now have global steel companies in an entirely new way than what there was in the past. While international steel trade has been around for a long time, the unprecedented trade pressures and disputes of the 1980s (Chapter 7) led to a major change and formation of joint ventures with foreign producers. This made for a somewhat disjointed and transitional stage for the industry. It was followed over the past decade by the complete international reorganization of the industry through mergers and acquisitions, along with many exiting companies. The new global steel companies are still very much a work in progress and will continue to evolve. For example, as discussed later in this chapter, the physical boundaries of the firm are an issue, both at the raw materials end and at the distribution end of the business.

The industry has seen a huge consolidation of steel companies driven by bankruptcies, mergers, and acquisitions. The list of companies itself tells the story. The most dramatic changes were within North American operations.

- Algoma Steel (assets bought by Essar Steel, India in April 2007)
- Arbed (merged with Aceralia and Usinor 2002 forming Arcelor)

- Arcelor (merged with Mittal forming ArcelorMittal)
- Bethlehem Steel Corporation (assets bought by ISG in 2003. ISG merged with Mittal, now ArcelorMittal)
- British Steel (merged with Koninklijke Hoogovens (NL) in 1999 to form Corus, now Tata Steel)
- Carnegie Steel Company sold to U.S. Steel
- Cockerill-Sambre (acquired by Usinor in 1998, which became part of Arcelor in 2002, now ArcelorMittal)
- Corus Group (acquired by Tata Steel in 2007)
- Dofasco in Hamilton, Ontario (acquired by Arcelor, now ArcelorMittal)
- Hoesch Stahl AG (acquired by ThyssenKrupp)
- Inland Steel Company (acquired by Ispat International, became Mittal, now ArcelorMittal)
- International Steel Group (merged with Mittal, now ArcelorMittal)
- Jones and Laughlin Steel Company (acquired by Ling-Temco-Vought, renamed LTV Steel, acquired by ISG)
- Koninklijke Hoogovens (merged with British Steel (UK) in 1999 to form Corus, now Tata Steel)
- Krupp (merged with Thyssen to form ThyssenKrupp in 1999)
- Lackawanna Steel Company (acquired by Bethlehem Steel in 1922, plants closed in 1982)
- Laiwu Steel (merged into Shandong Iron and Steel Group)
- Lone Star Steel Company (acquired by U.S. Steel in 2007)
- Mittal Steel Company (merged with Arcelor, forming ArcelorMittal)
- National Steel Corporation (acquired by U.S. Steel in 2003)
- Northwestern Steel and Wire closed in 2001. Later partially reopened by Leggett and Platt, as Sterling Steel Company LLC
- Republic Steel (merged into LTV Steel, acquired by ISG, merged with Mittal, now ArcelorMittal)
- Rouge Steel (formerly owned by Ford Motor Corporation) acquired by Severstal in 2004
- Steel Company of Wales (absorbed into British Steel in 1967)

- Stelco (acquired by U.S. Steel in 2007)
- Thyssen (merged with Krupp to form ThyssenKrupp in 1999)
- Weirton Steel (acquired by ISG, which merged with Mittal, now ArcelorMittal)
- Youngstown Sheet and Tube (acquired by ISG, which merged with Mittal, now ArcelorMittal)

Source: Wikipedia.

The result of the wave of consolidation was to establish huge international steel companies. The following is the list of the top 12 steel-producing companies in the world by capacity as of 2010. As can be seen it is heavily weighted toward Asian countries.

Rank (2010)	Capacity (Mtns)	Company	Headquarters
1	98.2	ArcelorMittal	Luxembourg
2	52.9	Hebei Iron & Steel	China
3	37.0	Boasteel	China
4	36.6	Wuhan Iron & Steel	China
5	35.4	POSCO	South Korea
6	35.0	Nippon Steel	Japan
7	31.1	JFE	Japan
8	30.1	Jiangsu Shagang	China
9	25.8	Shougang	China
10	23.5	Tata Steel	India
11	23.2	Shangdong Iron & Steel	China
12	22.3	U.S. Steel	USA

Source: World Steel Association.

Of the top 12, all but two are in Asia. There is one European, ArcelorMittal and one North American company, U.S. Steel. They are also all integrated steel producers. The implication is clear; the future of the steel industry will be led by a relatively small number of integrated producers with a dominant presence in regional markets. This suggests a different outcome than the minimill-led steel innovation story that has dominated the steel news of the past 20 years.

However, looking at it from the perspective of the past 5 years, the trajectory of that development may be greatly impacted by forces largely outside of steel companies' control, that is, the trajectory of raw materials prices and development. As shown in the following table, even with this level of consolidation, globally, the level of concentration in the steel industry is dwarfed both by those in the industries upstream where it sources its raw material inputs and by its key downstream consuming industries such as automotive.

Steel, Auto, and Raw Materials: Levels of Concentration (%)

Industry	Percentage output of top 10 producers (%)
Steel	23
Iron Ore	96
Metallurgical Coal	85
Automotive	80

So, while steel is big, it is severely constrained when it comes to upstream and downstream pricing power. It is this that has made the dramatic and volatile prices changes in the price of raw materials so destabilizing for steel producers.

Who Are Some of the Key New Global Steel Companies?

In concentrated industries, it may make sense to work outward from the lead firms.

Who are some of the key new players?

U.S. Steel

U.S. Steel was the leading steel producer in the first half of the 20th century. It dominated the US market and was looked to around the world as the model of what a steel company should be. With the difficulties in the industry in the 1980s, the company diversified into other fields including energy and transportation. More recently, it restructured and regained its focus in steel, expanding into Eastern Europe and into Canada as part of the global consolidation in steel. It supplies the automotive and energy sectors. It is now organized around U.S. Steel operations for North America and U.S. Steel Europe. The company manufactures a wide range of value-added steel sheet and tubular products for the automotive, appliance, container, industrial machinery, construction, and oil and gas industries. U.S. Steel is a leader in both process and product technology. The company has three research and development facilities.

Nucor

Nucor, originally the Nuclear Corporation of America, appointed Ken Iverson as president in 1965 and he changed the industry with the new electric arc furnace (EAF) technology, a staunch antiunion policy, and introduction of the first minimill to produce flat rolled products in 1987. Nucor is headquartered in Raleigh with major production operations across the United States and in Canada. It is a global player because it has been for 30 years the reference case for minimill technology developments. Because EAF steelmakers produce almost entirely for regional markets, all of its operations are in the NAFTA region. Through acquisitions and joint ventures, the company eventually grew to become the second largest American steel producer. Because of the nature of the electrical furnace technology, Nucor has adapted to fluctuations in the market more by reducing volumes in downturns rather than by reducing prices. In addition to its head start on the technology side, much of its success is the result of a highly motivated, team-based workforce with high profitability bonus compensation.

ArcelorMittal

ArcelorMittal is the world's largest steel company, headquartered in London and Luxembourg, with operations in more than 60 countries. It is present in all major global steel markets, including automotive, construction, household appliances, and packaging. ArcelorMittal is a leader in R&D and technology development. The company has its own supplies of raw materials accounting for about 60% of requirements and has its own distribution networks. ArcelorMittal was formed in 2006 by the merger of Arcelor of Luxemburg and Mittal. The latter was originally from India but later consolidated Central and Eastern European operations, and then took over a series of distressed North American steel companies. ArcelorMittal continues its strategic policy of horizontal consolidation. More recently, it has been seeking to expand operations in China and India. Construction and infrastructure are the leading steel sectors in emerging market countries and the company's product market and facilities configuration have been similarly focused.

Posco

Posco is headquartered in Seoul, Korea. It was incorporated in 1968 with close ties to the government who had the objective of making the country self-sufficient in iron and steel. It has expanded and formed joint ventures in China, India, Vietnam, Mexico, and Brazil. Most observers regard it as the most technically advanced steelmaker in the world. It has major expansion plans in Asia and in Brazil based on its indigenous technology advantage. It also has an announced policy of displacing the Steel Service Centers in order to get closer to the knowledge base of its customers. It has developed the Finex process for iron-making, which may eventually replace the traditional blast furnace with significant reductions in production costs and fewer environmental impacts.

Gerdau

Gerdau Group is headquartered in Brazil. It is the world's 14th largest steelmaker and the largest producer of long products in the Americas. Gerdau Ameristeel is the fourth largest overall steel company and the second largest minimill steel producer in North America. The company's products are used in a variety of industries, including construction, automotive, mining and electrical transmission. It has a vertically integrated network of minimills, scrap recycling, and downstream facilities. Its products are generally sold to steel service centers and fabricators and a minor share going to original equipment manufacturers.

ThyssenKrupp AG

ThyssenKrupp is the largest European steelmaker. It is a result of a merger between other leading German steel companies. The company is a technical leader in metallurgical engineering, process technology, and new application development in the automotive, construction, and energy industries. It is a global leader in advanced steel product development with engineering applications especially supporting the German auto industry, which leads the world in automotive design and tooling.

Tenaris

Tenaris is headquartered in Buenos Aries, Argentina. It is a leading supplier of tubular products and related services for the world's energy industry. The company's principal products include casing, tubing, line pipe, and mechanical and structural pipes. It produces seamless products for high-pressure, high-stress applications, and welded pipe for more standard environments. Tenaris expanded in the past decade into the United States principally through the acquisition of Maverick Steel and Hydril. It operates Tenaris University, which gathers and codifies the knowledge and best practices within the company's operations for both salaried and hourly employees.

Baosteel

Baosteel is the one of largest steel company in China. It is state owned and is headquartered in Shanghai. With the economic reforms in 1978, the government decided it needed a major steel capability located close to Shanghai. It incorporated the latest technology then available, principally from Japan. As other Chinese mills came on stream, Baosteel decided to diversify into exports. Baosteel also absorbed several other money-losing state-owned steel companies. It later formed a joint venture with Thyssen Krupp of Germany. It is rightly regarded as the symbol of the rise of Chinese steel. It has been expanding through co-ventures in Brazilian iron ore developments.

Tata Steel

Tata Steel is the largest private sector steel company in India. It is part of the Tata Group comprising a broadly based enterprise with interests from financial services to chemicals to automobiles. It recently acquired Corus Steel of the UK and has announced the intention to expand to 100 million tons of capacity, half by building greenfield facilities and half by acquisitions. It exemplifies the rise of India as a global steel power. India has plans to add more steel capacity in the coming decade than does China.

How Steel Companies Used to Operate: The U.S. Steel Model

As indicated, steel companies used to be highly integrated vertical hierarchies managing the process from iron ore in the ground to delivery of a steel coil to a stamper's loading dock. They were present at the birth of the modern business corporation. The first billion dollar corporation ever formed was U.S. Steel in 1901. Given the scale of capital and operations, vertical integration was seen to be absolutely necessary, both to reduce transaction costs and to control production flows.

Steel companies were a huge story in the economy. The steel industry had the largest stock of plant and equipment in the US economy in the 1950s and from 1929 to 1958 accounted for the largest single component of GDP of any domestic industry. Even later, in most years it retained its place in the top three contributors to GDP till the mid-1970s.

From 1900 to 1960, the U.S. steel industry had an entrenched oligopolistic structure. Prices were based on a mark up over costs[1] and market demand was closely coordinated with industry production capability. The large integrated steel producers competed for market share to maximize internal utilization rates.

However, steel companies lagged other industries in modernizing their organizational structures and cultures. In 1962, the great business historian Alfred Chandler observed that the steel industry was virtually alone in staying with the centralized, hierarchical form of organization when all other major industries by the 1940s had moved to some version of the multidivisional corporation pioneered by GM and Dupont.[2] The GM model not only allowed it to produce different cars for different market segments—Chevrolet, Pontiac, Oldsmobile, Cadillac—it also made for a more diversified and innovative company.[3]

The steel industry was not alone in resisting the new multidivisional structure. The traditional approach was also kept in other metal and materials industries. Copper and nickel companies paid the least attention to the new management philosophy. Steel companies devoted more thought to the lines of authority and communication, in almost every case, bringing increasing centralization.

Steel had a long history of integrated activity, from refining of ores, converting iron to steel, and the fabrication of semifinished products all being done in the same plant. They produced a standardized set of products for a large number of customers, often accumulating large volumes of inventory in advance of orders. The higher volume of goods, large number of customers, and integrated production have made scheduling of operations and coordination functions much more complicated than the copper industry for instance. The production of standardized items in high volume ahead of orders also created more of a need to forecast and analyze future market trends in order to integrate the various parts of the enterprise.

After its initial consolidation, in its first operating period, U.S. Steel functioned as a somewhat loose federation of units with a small holding company office in New York. The major reorganization came between 1929 and 1937 under President Myron Taylor. It included building a large general staff, including a centralized Production Planning Department. The functioning departments were left in Pittsburgh including operations, sales, industrial relations, research and metallurgy, and traffic, each with its own vice-president. This was a step toward a multidivisional organization, but in 1950 U.S. Steel reverted to a renewed centralism. Central Operations took over all steelmaking. All activities under it were departmentalized along functional lines. In Chandler's view, U.S. Steel after 1950 became more like that of DuPont prior to 1921. Other steel companies followed the U.S. Steel pattern over the 1950s.

As stated, U.S. Steel in the mid-1960s had a culture replicated by most the big steel companies. Production men ran the company. Their only management metric was tonnage of steel produced. Quality and customer service were secondary. The organization was inbred, centralized, and autocratic. The main focus was not on products or markets.

It was 50 years later that steel companies began to experiment with multidivisional organization—separate operating companies for steel production facilities, distribution, transportation, and raw material inputs. Meanwhile, the world had moved on to matrix-style organizations that were much more flexible and emphasized cross-functional coordination and work teams.

There was a major North American exception to this direction, the Canadian steel producer Dofasco, which in the 1990s would become the most profitable integrated steelmaker in North America. When it became a fully integrated steel company under its founder Frank Sherman in the mid-1950s, he consciously decided not to go the route of U.S. Steel and the established form of organizational development. Instead, he developed an early form of the matrix organization. The pioneers in this area were ITT in the US and Panasonic in Japan. Dofasco's success in innovation in the 1990s flowed in no small measure from this different trajectory of organizational development taken 40 years earlier. It was also nonunion and built an inclusive workplace culture. The latter was ultimately probably more important in the long run than whether or not it had a union. It allowed the company to operate in a fundamentally different way.

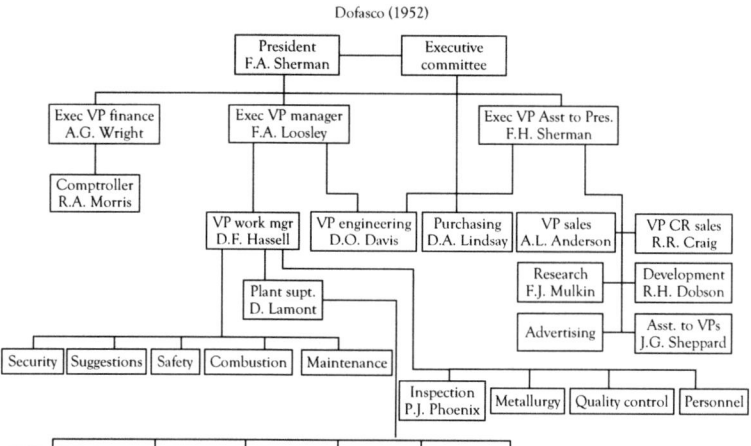

Dofasco Steel Organization Chart (1952)

However, across the industry, the narrow organizational form dominated. As Chandler stated:

> ...the production and marketing facilities for ore and semi-finished
> and even finished products cannot be easily transferred to
> the production and marketing of new and different products.
> Therefore, the strategy of diversification with its inevitable
> increasing long range decision-making has not been a tempting
> one in the expansion of the steel industry. (337)

Chandler's theory of the modern corporation was based on a linkage between skills and organizational structure. He identified steel companies as typifying the *functionally departmentalized* organization as examples of early 20th-century capitalism. It divided the organizations along discrete tasks, that is, manufacturing, sales, finance, and centralized decision making. He argued that for the future, the *multidivisional form* was the most efficient organizational response to the increased complexity of expanded administration. Decision-making authority would be re-distributed to various functional operating units. Middle management would coordinate product flow from production through distribution, while top management defined and allocates unit responsibilities, monitored and controlled their activities, and plans and allocates corporate resources for future activities.

Chandler had identified what he thought to be the "logic of managerial enterprise," the dynamic of growth and competition of modern industrial capitalism. Such large, oligopolistic industries would be efficient because competition for market share and profits sharpened the skills of senior management. The key was the scale and scope of the modern industrial enterprise.[4] Scale refers to the cost advantages of large plants; scope meant the use of many of the same raw materials to make a variety of different products. Steel companies therefore stood at the head waters of managing operations in the modern industrial enterprise.

Such exemplary, focused corporate headquarters skills, however, did not spare U.S. Steel from making perhaps the biggest strategic error in business history in the mid-1950s when it invested backward instead of forward in steel technology. This also happened at the exact time the corporation was recentralizing.

In the 1950s, the American industry was the preeminent leader in the world. U.S. Steel alone had more capacity than Britain, Germany, and Japan combined. But by that time, the capacity built up during World War II had to be renewed and new capacity had to be built to meet the demands of the postwar consumer economy. U.S. Steel and the industry was awash in cash. Unfortunately, the company, followed by its fellow American steel companies, decided to re-invest backward in the old technology of the open hearth furnace, notwithstanding that the new technology of the basic oxygen furnace and the continuous caster were already installed in other parts of the world.

The drivers of the decision were the traditional culture of integrated steel companies, the tax code, and the regulatory environment.

The U.S. Steel executives were in part the victims of accounting rules. As depreciation is normally spread over 20 years, 1955 investments could not be fully recouped till 1975. Retirement of production facilities early would eliminate millions in tax deductions and undermine operating profits. Similarly, for raw material assets, they did not invest in new higher quality iron ore and coal resources because they would have to write off earlier investments in existing mines.[5]

The executives who made the decision were also constrained by their personal histories. Almost all of the top executives were former open hearth engineers.[6] It was the only production technology that they knew. What it meant, however, was that their personal pasts compromised their company's future. By the 1960s and 1970s, the US producers were fundamentally uncompetitive with Japanese and European producers using the new technologies. The following 30-year slide in the US industry would really only end with the merger and acquisition movement of the past decade.

In the early 1980s, U.S. Steel produced a ton of raw steel in 11 man hours per ton. By the mid-1990s it produced a ton of raw steel in 4 man hours. The massive downsizing of U.S. Steel in the 1980s resulted in a company that was basically an exclusive producer of flat-rolled steel products. Legendary U.S. Steel mills were closed—Homestead, Duquesne Works, and McKeesport. U.S. Steel was frog marched into catching up with the revolution in continuous casting that the Japanese steel industry had led in the 1970s. Capacity and employment were cut in half. The number of mills was reduced from 15 to 5.

In the end, U.S. Steel, or USX as it had come to be called, was focused and profitable. It retained a physical advantage within the high end of the flat-rolled market over its new minimill rivals. Complex processes in the melt chemistry produce an irreducible surface quality advantage for sheet made from iron ore over similar products made from scrap that have inevitable impurities. The physical properties required for ultralow carbon steel, mainly formability, cannot be matched by scrap-based production. This became the last and most highly valued market niche for the integrated producers, of which U.S. Steel was the clear leader in the United States.

Managing Steel Companies in the New Global Environment

What is the impact of globalization on management of steel operations and what are the likely future directions?

The most immediate, practical impact has been significant cost savings, reportedly in the 10–15% range, through global benchmarking of best practices within the new global steel management system. These range from technical operating, engineering and procurement practices, to human resource policies. It has also been accompanied by circulation of technical talent both inward and outward bound between domestic and international operations. This has now become embedded in standard management practices and procedures, including regular monthly and quarterly meetings and conference calls, along with formal annual reporting and quantitative measurement. Management practices have become much more systematic.

The second and more important long-term factor is the change in access to capital and technology. Private sector steel companies have always been capital constrained. Access to capital has significantly increased and decision making on capital expenditures has been significantly speeded up say managers who worked under the old system as well as the new.

The following are variations on the same theme from several different steel companies:

There is greater access to capital but it is competitive. Things like environmental regulations are a factor. Best practices have reduced overheads. There is more leverage with suppliers. We have become a more sophisticated company.

Capital access is much better. We have dramatically improved our scrap business, a new reheat furnace, new cold bar sphere, other upgrades. This started with the previous ownership but accelerated with the new owners. But there is competition within the organization. We have to get the benchmark returns.

Steel Executives

Technology transfer has also improved under globalization. The industry in the previous 20 years had become increasingly reliant on international technology licensing and transfer to try and stay abreast of the latest and greatest in steel technology and applications. However, those who worked with technology licensing are of the view that the vendors never provided the absolute best-of-breed solutions.

We only ever got about 2/3 of the knowledge. We never got the best talent and latest stuff. We would get 590 but meanwhile the other companies had moved on to 780. We never got what we contracted and paid for.

There is fuller access now.

Steel Executive

As indicated in the quotes, under licensing, the local managers were always trying to catch up and never believed that they were getting all that they paid for. Now they believe that they get full access to the complete range of leading-edge technology from around the world.

However, this improved and expanded access to capital and technology comes at a price. The managements of operations are in an intense competition with other facilities trying to get their projects accepted and funded by head offices.

Steel Executives Objectives

Recent research by the author indicates that CEOs of steel companies currently identify five long-term operational and business objectives:[7]

1. Zero accidents in an inherently dangerous production environment.
2. Zero defects in the face of relentless pressure from quality-conscious customers.
3. 100% reliability for on-time delivery into lean production environments.
4. Ongoing productivity gains through continuous improvement.
5. Penetrating new markets with innovative products, while at the same time achieving constant cost reductions.

This agenda of challenges for managers frames the larger issues discussed in this book. To understand steel management, we have to begin with the global context. New global steel companies manage differently from their classic steel company predecessors.

New Global Steel Benchmarking

A major impact of the globalization of ownership in the steel industry has been the new and rigorous application of international benchmarking norms. Benchmarking applies to all aspects of production and management, including human resources management performance. For local management, achieving benchmarks is critical to compensation as well as attracting new investment and new technology. In the long run, achieving benchmarks is essential to the continued viability of a production facility. They have also become a key driver of innovation at the plant level, including organizational development strategies.

Traditional Steel Industry Performance Benchmarking

Traditional steel industry performance metrics were fairly simple. At the operations level, performance was measured by tonnage volume and capacity utilization. By the 1970s, additional economic metrics

Traditional Comparative Data for Steel Industry Production Costs

| | Inputs (cost per tonne) | | | | | | Cost structure (per tonne of steel) | | | | | |
| | Iron ore | | Coke | | Wages ($/h) | | Pig iron | Scrap & DRI | Raw material | Other | Liquid steel | Slab/ billet |
	2003	2005*	2003	2005	2003	2005	2005	2005	2005	2005	2005	2005
Integrated mills												
USA #1a.........	$40	$57	$129	$206	$48	$50	$184	$43	$227	$72	$320	$353
USA #1b.........	40	57	140	228	48	50	190	42	232	72	327	368
USA #1c.........	44	65	144	280	48	50	201	57	258	72	355	389
Europe #4a......	41	81	133	243	45	48	228	41	269	73	369	403
Europe #4b......	41	81	144	250	45	48	230	40	270	73	371	405
China #7a.......	53	83	93	153	1.96	2.50	170	53	231	53	308	332
China #7b.......	87	133	93	153	1.96	2.50	244	55	299	53	382	409
China #7c.......	37	78	93	153	1.96	2.50	173	53	227	53	303	327

(Continued)

Traditional Comparative Data for Steel Industry Production Costs—(Continued)

	Inputs (cost per tonne)						Cost structure (per tonne of steel)					
	Iron ore		Coke		Wages ($/h)		Pig iron	Scrap & DRI	Raw material	Other	Liquid steel	Slab/billet
	2003	2005*	2003	2005	2003	2005	2005	2005	2005	2005	2005	2005
Mini-sheet mills												
USA #2............	na	na	na	na	45	48	31	195	226	81	327	348
Europe #5.........	na	na	na	na	45	48	64	185	249	80	351	373
China #8.........	na	na	na	na	1.96	2.50	0	274	275	69	369	388
Mini-bar mills												
USA #3............	na	na	na	na	43	46	0	180	187	88	292	307
Europe #6.........	na	na	na	na	43	45	0	196	201	87	306	321
China #9.........	na	na	na	na	1.96	2.50	0	273	275	68	368	391
China #10........	87	133	na	na	1.96	2.50	292	0	292	53	374	398

Source: Steel Industry Update #198 July 2005.
Notes: *As of May, 2005.
+Before yield loss.

were added, utilizing costing expressed as Man Hours Per Ton (MHPT) at the raw steel, hot-rolled, and cold-rolled stages of steel production. The latter became common measures for productivity comparisons between national steel industries. An example comparison table is given on the previous page.

Such metrics are still used; however, a new set of measures and practices have been developed for management of multiple operations across nations and regions.

Reference Mill Labor Productivity Benchmarking

The new companies have developed different sets of metrics for multinational operations based on reference mills. How they individually implement this varies considerably by company. In the following discussion, Ameristeel is a US-based steel company with operations in North America and Europe. Eurofer is a large European-based steel producer with multiple steelmaking operations across all product markets in North America and around the world. Metallos is a South American-based steel maker primarily focused on the energy industry in all major markets.

Ameristeel Case

Ameristeel bases its mill metrics on a single reference mill, a key flat-rolled steel facility in the United States. All other facilities, in America and in Europe are based on this one reference point. It is a labor time metric based on the full time, direct labor costs expressed in terms of the number of employees per production department and at each stage of the production process. It is basically a head count metric.

The facility work force is segmented into Active employees, Laid-Off, Active Retirement-Eligible, and Collective Agreement-based. The labor productivity cost is calculated based on the Active Employee totals and number of employees by Department. The units are numbers of Employees. For reasons of confidentiality, only the categories not the actual figures are included.

Departments	Primary steelmaking operations only (No. of employees)	Hot-rolled coils (primary and zinc line operations) (No. of employees)	Cold-rolled (primary and cold rolled operations) (No. of employees)	Total (No. of employees)
Operating				
Coke				
By-products				
Blast Furnace				
Raw Materials				
Steelmaking				
Conditioning				
Pickle Line				
Ops Services				
Bloom & Bar				
Total Operating				
Maintenance				
CO/BF Maintenance				
Steel Making Maintenance				
CRC/HRC Maintenance				
Ops Services Maintenance				
B&B Maintenance				
Plant 24/7 Maintenance				
Total Maintenance				

The labor cost-based productivity comparisons are then used to benchmark domestic and international plants in comparison to the Reference Plant on a monthly basis. The results are then aggregated into the relevant product line, in this case flat-rolled steel products.

Facility	4Q09 (No. of employees)	1Q10 (No. of employees)	2Q10 (No. of employees)
Reference Mill			
Mill 1			
Mill 2			
Mill 3			
Flat-Rolled Total			

Eurofer Case

Eurofer has a much more ambitious set of performance benchmarks. They operate in many countries and identify three reference mills for flat-rolled steel, one each in Europe, North America, and South America. The performance and practices of these three provide the template for the others in the group.

The Eurofer Benchmarking Program Objectives include: Health & Safety, Organizational Effectiveness, Repair, Maintenance & Capex, Energy and Other Costs, Quality, Environment Improvement Programs.

Major contributors to differentiation between facilities include asset configuration: size, age, volume of facilities; operating practices: sinter versus pellets, etc.; and, local plant factors: wage levels, labor standards, and raw material logistics. Among the measurement challenges for these reference mills are ever-changing exchange rates, product mix, and employment practices, which can cause significant shifts in the ranking as shown in the examples below.

Benchmarking Metrics

	Mill 1	Mill 2	Mill 3
Total #FTE (own, contracting, overtime)	No. of Employees	No. of Employees	No. of Employees
Ratio own FTE/total FTE	% ratio	% ratio	% ratio
TCOE	$ per year	$ per year	$ per year
Total contractor cost	$ per year	$ per year	$ per year
% cost employees in total contractor cost	% ratio	% ratio	% ratio
Compare avg cost own fte	% ratio	% ratio	% ratio
Avg cost contractor/cost own fte	% ratio	% ratio	% ratio
# own blue collars/total # FTE	% ratio	% ratio	% ratio

A surprising finding is that contracting out is often more expensive than in-house costs. Contracting out is historically cheaper in South America but exchange rate changes are rapidly eroding this advantage. The cost metrics are then used to rank relative performance of the Reference Mills across global operations. As can be seen below, places in the ranking tables shift at various times between the different regional mills.

Comparative Mill Productivity Metrics 1: Ranking Shipments/FTE own + overtime + contractors		
2008	2009	2010
Mill 1	Mill 2	Mill 2
Mill 2	Mill 1	Mill 1
Mill 3	Mill 3	Mill 3

Comparative Mill Productivity Metrics 2: Ranking HRC/FTE own + overtime + contractors		
2008	2009	2010
Mill 1	Mill 1	Mill 1
Mill 3	Mill 2	Mill 2
Mill 2	Mill 3	Mill 3

Comparative Mill Productivity Metrics 3: Ranking Employment cost/tons shipped		
2008	2009	2010
Mill 1	Mill 1, Mill 3	Mill 1
Mill 3		Mill 3
Mill 2	Mill 2	Mill 2

Productivity Metrics 4: Comparative Mill Employment cost/HRC: Ranking		
2008	2009	2010
Mill 2	Mill 3	Mill 2
Mill 3	Mill 2, Mill 1	Mill 3
Mill 1		Mill 1

To add complexity and give more granularity to management metrics, Eurofer is moving to asset-based measurements. Within the reference mill set, the facilities have been disaggregated into asset groupings by department and function.

Productivity Mill Metrics: Asset-based: No. of Employees

Asset area	Mill 1 No. of employees	Mill 2 No. of employees	Mill 3 No. of employees
Primary			
Finishing			
Maintenance			
Executive Team			
HR			
Health & Safety			
Infrastructure			
Support Services			
Metallurgy & Quality			
Continuous Improvement			
Engineering			
Environment			
Process Automation & Modeling			
IS/IT			
Purchasing			
Warehouse Operations			
Customer Service			
Order Fulfillment/ Production Planning			
Finance/Controller			
Central Warehouse/ Plant Stores			
Utilities			
Sales			
Total	No. of Employees	No. of Employees	No. of Employees

Eurofer also uses their three reference mills to do on-going experiments. For instance, Mill 1 is used to benchmark Maintenance practices. Mill 2 is the reference for Contracting Out practices. Mill 3 is being used for Knowledge Management practices for the production work force. Within this again, there are developing metrics for Human Resource functions and organizational effectiveness and innovation.

Steel Benchmarking and Time Metrics

In addition to these labor-based productivity metrics, benchmarking is also being applied to time utilization in new and finely grained ways. Steel, like all continuous, capital-intensive operations have always used the language of Up Time and Down Time, because profitability only really begins at about 80–83% of capacity utilization. More recently, with the incredible pressures of cost and delivery, things have been taken to a new level. Stoppages are the variable factors.

Metallos Case

Metallos has what appears to be the most top-down engineering-driven metrics for capacity utilization and "up time." Because the company produces products for the global energy industry with central supply contracts, their objective is a single industrial system standardized across operations around the world, to guarantee product technical specifications and quality. It is particularly directed to issues of time metrics for continuous production.

Metallos' methodology uses Calendar Time, Possible Time, Available Time, and Effective Time to generate a new set of ratios. Calendar Time is the time included in the period under analysis, for example, 24 days, 744 h per 31 day month, 365 day years. Structural Stoppages are times when the line does not have the crew to operate for structural reasons: holidays, union activities, rest periods, nonscheduled shifts, production suspensions, unscheduled maintenance, and vacations. Possible Time is the time the line is fully prepared to operate. Stoppages include scheduled repairs, change overs, peak energy periods, previous stoppages and subsequent startups, lack of demand, lack of energy, new technology installations, raw material shortages, logistics interruptions, weather, and preventive work. Available Time is Possible Time minus Stoppages. Effective Time is the time the line is actually running.

These definitions give rise to the following calculations:

$$Calendar\ Utilization = \frac{Effective\ Time}{Calendar\ Time}$$

$$Possible\ Utilization = \frac{Effective\ Time}{Possible\ Time}$$

$$Available\ Utilization = \frac{Effective\ Time}{Available\ Time}$$

$$Cycle\ Time\ Ratio = \frac{Standard\ Effective\ Time}{Actual\ Effective\ Time}$$

These ratios are then applied to benchmarking matrices within and between facilities as shown in the following table.

Time-Based Productivity Metrics

	Actual (No. of hours)	Objective (No. of hours)	Var vs. obj % ratio
Calendar Time			
Possible Time			
Available Time			
Operations Delays			
Nonoperations Delays			
Total Delays			
Calculations	% Ratio	% Ratio	% Ratio
Utilization (Effec/Avail)			
Utilization (Effec/Poss)			
Utilization (Effec/Calendar)			
Efficiency (Perf/Util)			
Productivity (tons/effect hr)			
Productivity (SRM pcs/effect hr)			

Taken together, these new and sophisticated labor productivity and time metrics are being integrated with the conventional financial metrics in assessing steel company operations. They directly impact executive compensation and capital allocation decisions. This is an entirely new way of managing steel operations from the simple internal hierarchies of the past.

Global Supply Chains and the Reverticalization of Steel

About 70% of steel company costs are related to raw material inputs. Therefore, how a steel company manages its raw material supply chains is critical to its survival. To illustrate, we will look at the case of metallurgical or "met" coal. Met coal stands at the most upstream of all of these production activities. Even the seams of coal in the ground at the mine have critical properties for the whole downstream processing and product development.

Coal is seen by many as the ultimate industrial commodity. It is black, dirty, and bulky. Much like steel itself, it is a classic of the old, industrial economy. However, coal is now in transition as a value-added resource input to advanced steel manufacturing. In academic language, this has meant a classic shift from commodity production to its full integration into the value chains of the new steel industry.

The seams of coal at the mine have critical chemical and physical properties. Once the coal is out of the ground its properties cannot be changed. It can only be blended with other grades. The extracted coal is separated and managed in terms of these elemental properties. At the mine operations level, these coals are blended into standard mixes and are labeled "components" for the subsequent processes. In global steel, reliance on seaborne coal is the new standard practice. The coal is moved in unit trains to terminals on the coast. It is there that the components are combined into "products" to match the specific grades steel-producing customers have specified as the "recipes" they need for steelmaking downstream. This takes place literally as the coal is loaded on to the vessels. Products may even be segregated within specific holds on the bulk carriers. A typical coal producer may produce eight to ten components, which are later blended into 20 different products.

The most desirable product in global metallurgical coal is HCC (Hard Carbon Coal) preferably with medium volatility. The prime grade HCC coals are scarce and much desired because they ultimately have major impacts on the efficiency of the steelmaking units in the mills and thereby affect productivity, profitability, and product quality for the steel companies. The highest grade HCC is critical in the Blast Furnace because of its chemical properties and also because it is strong enough to keep the sides of the furnace from caving in. This is most important with the newest generation of large-diameter blast furnaces developed originally by the Japanese, then extended by the Koreans and now being installed in the new mills in China and India.

There are only three established sources of this HCC with premium qualities: Australia, Western Canada, and the USA. The latter's sources are rapidly dwindling, leaving Australia and Canada as the lead competitors. The former have about 65% of the market and the latter about 35%. This is the heart of the HCC seaborne market. There are potential new sources in Mongolia and Mozambique, but significant political risks and infrastructure development challenges constrain their early entry into the international market place.

For coal companies, transportation and distribution costs are their highest fixed costs. National, country-specific railway freight rate determination by public agencies has become wild cards for price and development decisions. The Regulator is more concerned with system maintenance costs than optimization of the specific supply chains that undergird the whole of the system. In addition, there are coordination and access issues around the terminals and access in general at the ports. Getting the coal shipped has become as complicated as its extraction and grade determination.

The actors and regulators of the supply chain should step back and take a breath while they reconsider the interrelationships and long-term interests of the met coal producers and the optimization of the supply chain. To take one example: if unit costs of transportation can be reduced by optimizing performance through use of larger trains, better traffic coordination and more effective access to ports and terminals, then this feeds back directly into economics of mine operations. Reduced fixed costs for transportation will allow the mine operators to literally go deeper in their seam cuts, utilize larger mines with longer ore body life, improved economics for all levels of extraction, make the mines more productive, and employ more people for longer. It should probably result in higher not lower revenues for system operators, railways, and terminals in the long run.

Virtually, all industry observers give optimistic outlooks for metal-lurgical coal for the next 10–20 years. The fundamental driver is Chinese steel demand to support huge steel-intensive infrastructure projects. Demand especially for high-quality HCC coal will exceed supply for the foreseeable future.

Prices movements may be a different issue. Mining conferences and business publications have been filled with excessive exuberance about a China-led super cycle in metal prices that is expected to run for decades. Financial markets seem to have bought into the story. However, there are good reasons to be skeptical about such a one-way economic forecast. There is an important debate going on in resource economics about this theory, especially in the light of the setbacks in 2008–2009.

In summary, the counter-argument is that there is not a one-way up escalator in metal prices. Steel in particular has always been a cyclical market and that factor is still with us. Demand from China is a huge story but locational factors may account for at least another one-third of the movement in prices. The other issue is historical supply problems. We are still reaping the downside of inadequate investment in mine development from the 1990s. All these factors should be considered both with respect to coal and metal price forecasts in general as well as mine and supply chain performance if the potential of the industry in future years is to be realized.

Implementing supply chain management successfully leads to a new kind of competition on the global market where competition is no longer of one company versus another company but rather it takes on the form of one supply chain versus another supply chain.

A value chain is a chain of activities for a firm operating in a specific industry.[8] Products pass through all activities of the chain in order, and at each activity the product gains some value. The chain of activities gives the products more added value than the sum of added values of all activities.

Value chains are networks, and the academic literature[9] distinguishes three basic types: modular, relational, and captive.

The Coal Chain is clearly not modular like the electronic and auto chains with closely specified OEM–supplier linkages, where the lead firms design platforms with tight material and performance requirements, which are then turned over to the suppliers to design, produce, and validate. Coal is not a widget. Even though coal producers have adopted the language of "components" they are not modular components, like parts and subassemblies are in the auto industry for instance. A coal company is not a Tier 1 auto parts supplier.

Coal, like iron ore, was once a predominantly captive supplier. Steel companies owned their own mines. As discussed above, the pressures on steel companies in the 1980s and 1990s, particularly in North America, to cut costs and revive profits, led them to largely divest of their iron ore and coal holdings. This was followed in the 1990s and aggressively in the 2000s by consolidation of coal producers into their own separate globally integrated companies. Buyer power, which was traditionally held by the steel companies, has now been replaced by Supplier power, most dramatically in iron ore and less so but importantly in coal.

However, this has to be qualified by understanding how the actual transactions take place. As described above, coal right from the seam is specified according to the product strategy and even the physical configuration of furnace facilities at individual steel companies and even specific mills within those companies. Hence, the all-important role of specification of the "recipes" needed for coal within the production parameters of the different mills. For this reason, the best characterization of the Coal Chain is that it is relational. Coal is not simply sold on an open market where price is the only factor. Much more information is required as the basis of the interchange between coal producer and steel producer. This inherently makes the transaction relational.

While there is much current discussion and speculation about whether steel companies will revert to direct ownership of mines, there are good reasons to believe that the story will play out differently between iron ore and coal. Many observers and commentators equate iron ore and coal but there are critical differences. Iron ore mining is much more concentrated and there is no alternative for integrated steel producers. Coal mining is less concentrated and there are other options: pulverized coal injection (PCI), natural gas, directly reduce iron (DRI), etc.

For this reason, while there may well be ownership changes in coal, including steel companies buying interests or shared ownership, the fundamental character of the Coal Chain will by its nature remain relational.

These issues of raw material supply in steel raise important questions in current industrial economic theory. Value chains capture the process of sequential transformation from inputs through stages of transformation to outputs and through to distribution and final consumption. It is related to transactional cost economics and based on three criteria: complexity of transactions, ability to codify transactions, and capabilities of the supply base. The increased complexity of global supply chains is changing our perspective on steel company economics.

There has been a major change in restructuring and locational decisions for steel facilities and capabilities.

> *The fulcrum of global steel development is raw materials and energy. It rises and falls on the cost structure. This follows the other metals groups like globalization in aluminum.*
>
> *The key is Brazil. It has this incredible base in low cost, high purity iron ore.*
>
> Steel Consultant

The major restructuring of the industry in the 1990s saw steel companies looking to disaggregate their operations horizontally (i.e., focus on what you are good at). Now the industry has been challenged to reverticalize. As mentioned, the trigger was the impact on steel prices and the cost of raw materials accompanying China's emergence as the leading steel power in the world. The story is told by a group of international steel engineering consultants.

> *Twenty years ago the companies disaggregated (do what you are good at) then with the rise of China and shortages/rising prices for raw materials, there was a wholesale re-aggregation and consolidation in the industry. It has turned around 180 degrees.*
>
> Steel Consultant

Ownership of iron ore and control of costs became strategic. Companies have leveraged back to the iron ore stage. It is the big competitive advantage. It is where Brazil has such an advantage. It has the cost advantage in iron ore and plays it through to the slab stage.

Steel Consultant

The key variable in accessing the lowest possible raw material costs has raised Brazil's profile for investment decisions. As suggested, Brazil may actually be a more important steel company restructuring in the coming decade than China. Companies have focused new capacity investment on Brazil at least for raw material processing up to the slab stage.

China may not be such a big story because it has to import everything and has such bad infrastructure.

Steel Consultant

Company operations will be impacted by decisions to locate basic production facilities close to the cheapest raw material inputs. Finishing capacities can then be located wherever the end-user markets are located.

There is a New Steel Paradigm

The old view was that raw materials were ore and coal.

The new view is that raw materials include right up to the slab and even hot band stage.

The end user only comes in at the finishing stage. You locate that close to the market.

Brazil is positioned to play this strategically but isn't there yet.

Steel Consultant

The big guys with ownership and control of resources will be able to choose where their intermediate products are positioned.

Value added will be assessed at each step in the chain of manufacturing and assessed in terms of the capital required.

Steel Consultant

There are two significant implications of this shift. In the past, raw materials simply meant commodity grade iron ore and coal, but now steel companies regard raw materials as including everything up to the slab stage and perhaps even the hot-band stage. This may have dramatic implications for the nature and scale of North American steel facilities in the coming decade. The test case of this is being played out in the new Thyssen facility in Alabama where all the raw materials up to the slab stage will be imported then the finishing will be specialized to feed the local auto plants. The international location of raw materials versus domestic processing capacity can fundamentally change how steel companies operate in the future.

The Steel Company of the Future

Where does one find a model for the steel company of the future?

Boeing may be as a good an industrial corporation reference point as any for operations of the new networked manufacturer. It was and is one of the premium manufacturing companies in the world. Its past was anchored in a deep aeronautical engineering and machining culture with strong social ties to Puget Sound. In the 1990s, it faced persistent dissatisfaction from investors and financial markets for its inability to generate consistent strong earnings, even when orders were buoyant. In response, it acquired McDonnell Douglas. In retrospect, this was a tipping point for changing the culture of the company.[10]

Modularization has been an important feature of Boeing's production system ever since the divestiture decision in the 1950s forced it to spin off its engine division into what would become Pratt & Whitney.

However, with the development of the 787 Dreamliner, Boeing took on the new role of system integrator. Traditionally, it pursued a rigorous system of boundaries on key information in which its Tier 1 and Tier 2 suppliers had a directive to simply "build to print," that is, build a component to specifications as provided. In general, Tier 1 suppliers are the upperend producers of subassemblies, while Tier 2 producers supply the former with individual parts. Tier 3 producers such as stampers are the furthest upstream the supply chain.

Two factors caused the system to move in a new direction. The sheer costs to developing a new plane exceeded what even Boeing could afford. Therefore, it looked to its suppliers to co-fund development with the inevitable consequence that they acquired a greater share of production knowledge and the levers for incremental improvement. Also, as Boeing pursued foreign sales, particularly in Asian markets, it conceded more production control and related knowledge to foreign operations. For example, Mitsubishi is close to Boeing in terms of its proportion of the final manufacturing content and share of value added for the air frames "made" in Seattle.

In the past *How to Build a Commercial Airplane* was the proprietary bible for Boeing engineers, incorporating 50 years of learning. With the 787 Boeing "opened" the Book and provided suppliers with performance specifications for parts and components and collaboratively worked with them in the design and manufacturing of major components such as the wing, fuselage section, and wing box. With assemblers sharing more of the "secret sauce" with suppliers, incremental technology improvements are increasingly likely to be replicated across multiple aircraft platforms. This becomes a strategy of sustaining innovations by migrating value from assemblers to Tier 1 suppliers.

More than 90% of the 787 is outsourced to a global multitiered supplier network. Boeing has sought to commoditize most components manufacturing while retaining strategic assembly expertise and proprietary processes for manufacturing selected key items. The objective at the Boeing end is that value migrates to and stays with the systems integrator.

There are tensions and competing interests in even the most advanced supply chains. For instance, in the value-added of a car, about one-third of the cost is logistics. It is premature to assume that the mere presence of time–space shrinking technologies in transportation and communications have eliminated these problems. The problems of actually moving materials, components, and final products still confront the increased complexity and geographic extension of production networks.

The Boeing Systems Integrator model for global manufacturing is still facing growing pains. The company has changed its self definition from manufacturer to designer and assembler. Famously, the 787 Dreamliner has been designed and coordinated through a distributed computer system with final assembly in Seattle being reduced from 30–40 days to 3 days. Even the holiest of holies, the manufacturing of the wings, is now done by Mitsubishi in Japan. While the concept and digital architecture may be leading edge, the implementation and administration of the new systems is incredibly complicated and uncertain. The time delays and associated penalties on the Dreamliner are reported to be in excess of $10 billion.

Arcelor may be the closest analog in steel to Boeing, seeking to use financial and performance metrics to coordinate a new kind of global steel company: the systems integrator. Another reference point is the ambitious attempt, being led by Thyssen Krupp, to have a control room in Rotterdam doing real-time coordination of shipments of iron ore and slab production in Brazil to producing plants in Europe and America within detailed specs and just-in-time schedules of lean production manufacturing customers. It is reported that there are continuing operational issues in its implementation.

Systems Integrator is an appealing concept as the organizing principle for the steel company of the future. However, getting there will be even more challenging than it has been for Boeing. The complexity of upstream supply chains and the fact that the steel company is not the final user makes the situation both more complicated and riskier. Alternatively, steel companies may not move forward to system integration. They could move one step back and limit themselves to reintegration of steelmaking and raw materials production in house.

In summary, supply chains are contested economic terrain. At the end of the day, a steel company may not have the power position of an automotive or aircraft OEM to play the full system integrator role in the future.

CHAPTER 4

How the Steel Industry Operates

Many people view the steel industry as the equivalent of a medieval fortress beset with the economic equivalent of siege warfare. In fact, the industry has been and continues to be in constant motion. The steel industry underwent two major reorganizations at the national level in the 20th century. It is now undergoing a third at the global level in the early 21st century.

Previously, because steel was mostly a stay-at-home industry, the steel industry story was primarily told as a series of national industry narratives, perhaps now supplemented by the new global aspect. However, based on the most recent scholarship in the area, this is to miss the major point. This time it is different, but we have been here before. The postwar steel story, at least for the leading countries of the USA, Germany, and Japan, was from the beginning a story that was one of global restructuring around a central theme of how the industry was best organized. Therefore, the new world steel industry should be thought of as entering but a new chapter in an ongoing global restructuring storyline.

The Organization and Reorganization of Steel Industries

The birthing form of the steel industry in the early 20th century saw dominant national firms emerge along the lines of the U.S. Steel model. In all leading industrial countries, huge organizations grew up around a tight vertical integration, owning and controlling all of the inputs from the mines through steelmaking to processing and distribution. What this meant at the industry level was a system where one or two dominant firms controlled production and prices across the economy. Oligopoly, if not monopoly, was the order of the day. The so-called Pittsburgh Price system, in turn was the central pricing mechanism in support of this system. The price of all steel products would be set by U.S. Steel in Pittsburgh and those prices plus transportation costs would be applied in all local markets across the country. It was also a means to keep steel production centralized so that new and regional competitors could not and did not emerge.

It was actually the New Deal economic model of how industries should work. It quite consciously set out to organize industries into oligopolistic companies with competition taking place within well-regulated rules of the game set by government. This was seen as a change from the destructive competition that led to the Depression and would also provide the base for sustained economic growth based on high wages, mass consumption, and continual high employment. In steel it would survive for 50 years.

To follow the developments and stages in the industry, it is best seen in historical perspective. Therefore, the following chapter follows the major themes in historical order. Three developments gave the industry its modern shape: the imposition of the New Deal industry model on the steel industries of the USA, Germany, and Japan. The Japanese steel technical revolution. And, the Nucor minimill revolution.

Big Steel in the USA, Germany, and Japan

U.S. Steel personified Big Steel. It dominated the American industry and overshadowed the rest of the industry in the postwar period, producing the entire range of steel products. At the industry level, the primary actors were a limited set of integrated producers—Bethlehem, Inland, National—producing steel largely for a mass domestic market, engaging in oligopolistic competition and counter-balanced by active trade unions and regulatory government. For the first 30 years of the postwar period, it was largely successful though the seeds of its future implosion were in sight.

Within the industry, in the 1970s and 1980s, it was common parlance among steel industry managers to say that the hard-pressed American steel industry was being destroyed by Japanese and European imported steel from producers whose major competitive advantage was that their steel industries had been bombed out in World War II then completely rebuilt with new plants.

Recent scholarship has turned this story on its head. Gary Herrigel's historical account[1] puts a whole new perspective on postwar steel developments. The New Deal Steel model was not only a model for the USA, but it was also exported, imposed by the Occupying Powers in Germany and Japan though these industries and cultures retained critical features determining how it was implemented. The overall model was the same but it did not lead to convergence, it led to variety so that in the 1970s, the American New Deal Steel companies would come under severe pressure by imports arriving from its supposed offspring.

The leading steel industries of the interwar years in the USA, Germany, and Japan had a remarkably similar structure. In each case, there was a large, deeply integrated, diversified producer of relatively high-volume standardized steel products accounting for half of the total output of the industry. In America it was U.S. Steel; in Germany, Vestag; and in Japan, Yahata. Around these lead producers there were a larger number of smaller, more specialized companies accounting for the rest of the industry's domestic output. Exports played a relatively minimal role.

In 1901, the year that U.S. Steel was established, the Japanese government incorporated Yahata Steel as the flagship for the emerging Japanese steel industry. The government had tried in vain to persuade the existing Zaibatsu industrial groups to enter steel, but the latter viewed the risks of large startup costs to be too much in terms of the expected returns. Yahata therefore took the lead in integrated Japanese steel production, largely using German technology. The Zaibatsu gradually moved into specialized steel production for machinery, shipbuilding, engineering, and trade aligned with their other industrial conglomerate activities. This established the backbone of the bifurcated steel industry so common in the interwar years.

In Japan, as in Europe, steel industries dealing with the slowdown of the 1920s followed by the contraction of the 1930s, resorted to industry cartels. By collectively setting prices, they sought to stabilize the industry. But there were persistent problems about inclusion of all producers in all markets. In the Japanese case, participation in cartels, while originally private, became compulsory and mandated by the government under the Major Industries Control Law of 1931. These set boundaries on product markets, prices in the industry, sales quotas, allocation of orders, and so on.

However, cartelization was still not enough to stabilize the Japanese industry. So, in 1933, the Japan Steel Manufacturing Company Law was passed to privatize Yahata and merge five integrated steel companies into the Japan Steel company. This was the entity that would dominate the industry. In wartime conditions, the collective governance of the industry was tightened to guarantee that Japanese steel supported military objectives through the Japanese Steel Association.

The interwar German steel industry had a similar macrostructure, but ultimately with different relations to the state. The dominant producer was Vestag but unlike Yahata it was a private company, the outcome of a merger in 1926 of the four largest producers with both integrated steel and coal facilities. It accounted for over 50% of German production. Around this core there were a set of specialized producers, the Konzerne, who had strong linkages downstream to engineering, shipbuilding, and machinery industries. However, unlike Japan, these were also integrated producers. Vestag focused on standardized, high-volume markets, while the latter focused on more specialized product markets.

The German industry responded to the pressures of the 1920s and 1930s with cartel organizations. However, these were private associations and never resorted to government compulsion for membership compliance like the Japanese Control Association.

As war approached, the German steel industry insisted on retaining its presence and autonomy in private markets and accommodated military demands only as a secondary priority. The Nazis responded by state-sponsorship of their own steel capacity in the form of the huge Herman Goring Works (HGW), which expanded to become the second largest producer. Vestag and others were not above taking advantage of orders for the Nazi state, but they maintained the boundaries of the private firm. HGW was joined by several smaller opportunistic ventures with a closer relationship to the militaristic state from within the Konzerne membership. Overall, the German steel industry was not as closely aligned with the state as was the wartime industry in Japan.

However, as the Allies looked to postwar reconstruction, they used their occupation powers to reorganize the industries in German and Japan to more closely resemble the American New Deal Steel model. The postwar objective of the USA was to dismantle the centralized steel regimes which were seen as major contributors to the Nazi and Japanese militarist regimes. The monopolistic steel firms were disbanded along with their related cartel organizations. A decentralized oligopolistic, mass production regime was seen as a significant economic and political contributor to postwar democracy. The key factors were breaking linkages to the state, decentralizing into a set of oligopolistic firms, and instituting trade union collective bargaining into decision making in steel.

The irony was that this new structure in the context of German and Japanese labor market policies, local institutions, and business practices became a platform for new technology, for example, the continuous caster and a more innovative, entrepreneurial steel industry that would emerge over the 1950s and 1960s into a superior competitor in the steel trade wars to that of its US point of origin.

It was this New Deal version of the steel industry that started to unravel in the 1970s. Between 1980 and 1994, US integrated producers wrote off $15B in underperforming assets.

Overcapacity in the global integrated industry became a chronic problem by the 1980s. All industries had to restructure. The Europeans and Japanese did it by cartelization through which they organized adjustment at the industry level. The American mechanism was protectionism in which price relief was given with the explicit understanding that individual companies would rationalize operations. In both approaches, government was given a major role in guiding the restructuring process.

In 1984, the Steel Import Stabilization Act gave the US Trade Representative the authority to negotiate wide-ranging voluntary restraint agreements (VRAs) to dramatically limit steel imports from 20 countries. The steel companies were to use the revenues to reinvest in plants and worker training. However, it never quite happened that way. Later, even the Bush administration extended the system. At its best, this was a form of negotiated restructuring. The forum was principally through Chapter 11 bankruptcy proceedings.

Steel operations in Germany and Japan also underwent wholesale processes of consolidation and rationalization. Germany and Japan would each eventually be left with two consolidated steel producers. Between 1980 and 1986 in the US, 24 of 47 steel plants were closed. The Asian Financial Crisis of 1997–1998 subsequently took the floor out from under global flat-rolled steel prices and precipitated a further restructuring not only in the USA but also in Europe and Japan.

The boundaries of steel firms were redrawn in the process. Upstream activities such as iron ore and coal holdings were often sold off. Internal processes from finishing operations, casting, maintenance, mill services, and even sales and marketing often became contracted with outside entities.

Again at the industry level, there was an additional strategic shift. The industrial structure of the steel industry essentially was inverted. Under US leadership, from 1945 to 1974, the standard format had been a set of large integrated producers focused on standard product lines with large volumes and a second tier of smaller firms producing more specialized products. The pyramid inverted in the 30 years from 1974. A much reduced number of integrated producers focused on high value-added specialized products, while the second-tier minimills took over lower value-added, volume product lines. It should be added that the lower value-added markets were also being supplied by new Asian and BRIC (Brazil, Russia, India, and China) producers.

The U.S. steel industry and market did not disappear; in fact, it remained critical for domestic and foreign producers. At the time when many people were thinking it had disappeared, in fact it became ground zero for radical restructuring of steel operations and commercialization of leading steel technologies from around the world.

The Japanese Steel Revolution

The next major piece of the complex puzzle shaping how the steel industry operates is to understand the nature and impact of the Japanese steel industry in the postwar period.[2] Postwar Japan produced a different kind of steel, one that was directly linked to the Japanese quality revolution in manufacturing. The interface between steelmaking and manufacturing changed qualitatively. Major Japanese steel producers emerged in association with industrial groups (Keiretsu) that had their own auto companies and other engineering concerns.

Beyond the individual steel companies, there were trade associations, professional associations of engineers, and overseeing it all, the hand of government. The Ministry of International Trade and Industry (MITI) saw themselves as active players in the whole industrial development. This had a major impact on the direction and pace of change. For instance, the Japanese government negotiated site licenses for the newest technologies, including basic oxygen furnaces and continuous casters, thereby lowering the cost of technology acquisition for steel companies and standardizing the common technical platforms so that companies and engineers could accelerate down the learning curve. Specific technology issues are discussed in Chapter 9.

The Japanese miracle was remarkable for its ability to achieve international competitiveness in industries for which Japan did not have any natural competitive advantage. The steel industry was perhaps "the" leading case of Japanese success. There was no indigenous natural resource base to support the steel industry; however, it was able to creatively source raw materials thereby creating a competitive advantage, for example, locational advantage through its tidewater steel plants. Cooperation between industry and government was critical to this success. MITI played an important role, particularly in the early stages of development by coordinating expansion and development. Steel firms were required to submit their expansion plans to the government, and MITI helped ensure that the firms could borrow sufficient funds to carry out the plans. Priority for the loans was given to the firms that seemed best able to modernize, reinforcing the need to invest in new technology and equipment. The government persuaded private banks to make preferential loans to the steel industry at low interest rates and rearranged payment schedules for previous government loans. The government also gave favorable tax treatment to the steel industry, including accelerated depreciation rates, lower property taxes, and exemption from duties on imported machinery and equipment.

The result was that Japan rapidly became the technical leader in the world steel industry and ultracompetitive in costs. But it also produced an industry that had to export about 50% of its volume in order to be profitable. This became the lightning rod for steel trade disputes discussed in detail in Chapter 7. The imperfect outcome of the latter disputes is that there was an important transfer of technical expertise in the 1980s and 1990s between the Japanese and American industries through the somewhat forced embarking on joint ventures, particularly on the processing side of the industry such as galvanizing mills for auto steel.

The Nucor Revolution

The next game changer in how the steel industry operates came in the 1980s with the upstart US producer Nucor.

The rise of the minimill in the United States was not simply a steel production technology story. The underlying factor was the vulnerability of the traditional industry to imports during the 1960s. This is attributed to two key factors: the unreliability of the central price-setting mechanism in the industry and the loss of control over distribution.

At the heart of the price-setting mechanism was the collective bargaining relationship between the United Steelworkers and the major integrated producers (the story is told in detail in Chapter 8). The history was a series of long strikes, crucially in 1956, 1959, and 1969. But, even when strikes were avoided, the customer base in automotive and construction became wary about continuity of supply. Previously, imports played a marginal role, but in the 1960s the scene changed. Major steel consumers started to import steel to tide themselves over the strike period. But from 1969 onward, the imports never left after the labor settlement. By the 1970s, the level of imports reached 20% of U.S. steel consumption, despite a series of protectionist interventions.

None of this would have been possible, however, unless distribution and purchasing changed. In the first half of the 20th century, steel producers sold most of their product through their own distribution channels. From the 1950s, small steel distributors started to emerge as a new form of competition. They were usually regional firms who at first sold scrap and then second-tier steel products. During the strike periods, these small firms were encouraged to significantly increase their imports of steel. Eventually this became the major channel for imports to penetrate the U.S. steel market. The cumulative impact was that domestic integrated producers lost control of their markets.

Even more importantly, the rise of the independent steel distributor opened the way for small independent minimill producers to selectively enter traditional steel product markets, often on a regional basis and creeping up the value chain from low-value bar products, eventually into the lucrative flat-rolled markets.

With pressure from both the imports and minimills using the regional distributor market, by the 1980s it was arguable that there was no longer a national steel industry as that would have been understood in the first half of the century, or even in the 1960s. For instance, the most famous of the new mills, Nucor, the second largest U.S. steel producer by 2000, served regional markets from a set of small EAF plants.

Nucor began its life in steel production by producing low-end bar products primarily for the construction industry. It followed other minimills in pursuing the market for rod and wire products. By the mid-1980s this market was getting crowded and Nucor management bet the company on developing new technology, licensed from the Japanese and the Europeans. The key was to couple the highly efficient electric furnaces to new types of continuous casting machines. In a joint venture with Yamamoto of Japan, in 1988 they opened a new mill to produce wide flange beams, pilings, and heavy structural for construction applications. By producing a "near net shape" much closer to the shape of the final product than traditional methods, the new mill reduced costs and allowed Nucor both to break into a market at a new price point that traditional integrated mills could not match and to distinguish itself from other minimill producers. In 1989, Nucor gambled again on an untried technology for "thin slab" casting, a simplification of the process steps from hot metal to a rolled coil by reducing the number of stages from about 20 to 6. This allowed it to break into the flat-rolled market. The German engineering firm SMS Schloemann-Siegman had pitched the technology at numerous companies, large and small, but Nucor was the only taker. By the mid-1990s, Nucor was also producing cold-rolled and galvanized products. By the late 1990s, Nucor was building iron carbide capacity to reduce its vulnerability to scrap prices and supply. No one doubted that Nucor was the most innovative steel producer in North America in the last quarter of the 20th century. By early in the new decade it had overtaken U.S. Steel as the largest North American producer.

US Crude Steel Production by Process (% of Total Production)					
Year	1990	1995	2000	2005	2009
BOF	59.6	60.7	54.9	47.9	37.2
EAF	35.9	39.3	45.1	52.1	62.8

Source: World Steel Association, Steel Statistical Yearbook, various issues.

There was another underlying theme in the Nucor-EAF story, the liberation of steel company management's right to manage. The new firms were mostly nonunion or union elimination situations. In other cases, it involved marginalization of the union, particularly the union at the nonlocal level. There is no doubt that there has been an infusion of new management into steel companies, along with new management vision and style. However, it is also the case that the majority of the problems in the traditional integrated steel industry has been the creation and inheritance of entrenched and overly conservative management.

In summary, scrap-based producers make lower quality steel, but with much simpler and cheaper processes. Because the input is already steel, the minimills use electric arc furnaces (EAFs) to simply transform the steel into more usable forms. Lower value-added steel, such as rebar for construction, can be produced with steel scrap alone. As EAF producers have pushed up the value-added chain into flat-rolled products, they have had to add high-quality scrap plus other iron products without the impurities contained in scrap. This includes pig iron, processed iron ore pellets, or briquettes of iron. Capital and labor costs for minimills are significantly lower than for integrated producers; however, EAF firms are very sensitive to the cost and availability of scrap. In recent years, the impact of demand from China for scrap has generated huge problems of both price and availability for EAF steel firms. Dozens have fallen into bankruptcy.

1980–1990s' Continual Restructuring

As described elsewhere in this book, by the late 1950s and early 1960s, cracks were appearing in the Big Steel system, which by the 1980s unleashed a flood of disruptive change. Domestically, the rise of the minimill challenged the traditional industry in long products. By the late

1980s, the EAF producers were pushing into flat-rolled products.[3] The long products were lower value-added and the big steel producers could not compete with the lower cost base of the minimills in terms of both capital cost per unit of capacity and production costs in operations. At the international level, oligopolistic steel pricing was undermined as imports flooded into the US market from Japanese producers in particular. Part of this was dumping of steel but fundamentally, for the reasons discussed in Chapter 9, the new Japanese superefficient BOF–Continuous Caster configurations simply had productivity advantages that could not be matched, even if the steel was not dumped.

In the 1996 book *The Renaissance of American Steel*, the two champions were U.S. Steel and Nucor.[4] By 2006, both U.S. Steel and Nucor were the subjects of takeover speculation by the European steel multinational Arcelor. How had the steel world changed and why?

Minimills did not have a strong tradition of research and development. The reduction and the elimination of R&D capacity at virtually all the North American integrated producers produced a situation where innovation came either from joint ventures with offshore steel companies or from global equipment vendors. Innovation within the North American steel industry came from a limited number of consortia, mostly with government agency support.

Between 1949 and 1959, global steel capacity almost doubled. By the 1960s, production capacity exceeded demand and the US market became ground zero for imports. Dumping cases abounded. Inflation increased wage, energy and transportation costs by 40% while steel prices increased by 10%. The U.S. steel companies fought back against foreign competition by seeking tax relief. With changes to the US Tax Code, American steel companies were able to invest more in BOFs.

However, by the 1970s and into the 1980s, the US companies had the worst on all sides. Lower profits reduced their stock values. But tax concessions resulted in higher cash flow than production levels alone would have generated. Together this made them takeover targets. As a defensive strategy, the steel companies diversified into other industries, that is, U.S. Steel moving into oil and gas. The Accelerated Cost Recovery System of the Economic Recovery Tax Act of 1981 provided so generous tax write-offs for capital-intense industries that steel companies could not use them all and created a secondary market, that is, by selling depreciation allowances. By the late 1980s the companies had retired all of their OH furnaces but hugely reduced their overall steelmaking capacities. Imports and minimills took up the slack. Integrated steel companies became holding companies, with separate subsidiaries for flat products, bar, structural, and raw materials. The 1990s would see further spinoffs of these units. These in retrospect were simply tactical, survival actions that were the preliminaries to the great steel consolidation of the next decade. The process started earliest in the USA, but eventually spread across the world steel industry.

The Great Steel Industry Consolidation Movement

The list of steel company bankruptcies, mergers, and acquisitions has been presented previously. The birth of the new global steel industry, in everybody's measure, has turned on the global wave of consolidation beginning in the late 1990s and lasting for a decade. Why did it happen? What was the trigger and what was the tipping point?

The OECD Steel Committee has given a broad summary of the major and sometimes competing theories for the dramatic consolidation of the global industry in recent years (OECD DSTI/SU/SC(2007)3/REV1).

The Fixed-Cost Hypothesis

According to this view, the steel industry, whose firms have a high proportion of fixed costs to total costs, are prone to periods of harmful price competition during market downturns. During periods of falling demand, if steel firms scaled back production to equal marginal revenue to marginal cost, they would quickly suffer profit losses since fixed costs per unit of output would rise sharply as production fell. To lower their unit costs, steel producers were tempted to lower their prices, produce more, and gain market share. As most producers faced the same incentive structure, the market price would fall, steeply at times, in response to the growing supply surplus on the market. This would result in detrimental profit losses, a situation that steelmakers would try to avoid by combining their companies. Thus, greater consolidation is a way to reduce price volatility and achieve higher profits.

This argument appears plausible if steel production were highly concentrated geographically with little or no trade internationally, as was the case a century ago. Production restraint in order to boost prices would thus not attract significant competition from steel imports. Today, steel production is dispersed across all parts of the globe and some 40% of it is exported. To a certain extent, price divergences can be sustained because imported steel is not a perfect substitute for domestic steel. Steel consumers may prefer locally produced steel due to, for example, the relatively short time needed to deliver it to customer manufacturing plants, and thus be willing to pay a premium over imported steel.

Economies of Scale

Related to the fixed-cost hypothesis is the idea that steel industry consolidation takes place because steel firms strive to take advantage of economies of scale. In other words, they achieve lower unit costs through higher production. If economies of scale are to be achieved, smaller steel plants have to be replaced by larger plants. However, consolidation in the steel industry often occurs through the acquisition of additional plants, which does not generate economies of scale in production. Therefore, economies of scale, alone, do not seem to be an important explanation of consolidation.

Synergies

Even though consolidating firms may not benefit from economies of scale in production, by coordinating the assets, know-how, and management skills of the merging firms, the combined steel firm is more efficient and thus enjoys superior output/cost combinations. Thus, synergies require the sharing of merging companies' assets, which allow the combined company to produce as much or more for a given cost. In the recent large mergers, the synergies cited relate mostly to marketing and product development, R&D, and purchasing. Combined companies may benefit from lower raw material costs through greater negotiating power over suppliers, from managerial efficiencies that reduce corporate staffing needs, and lower costs of distributing steel if the various distribution systems can be integrated as well. Such synergies can be significant. In the case of Mittal

Steel's acquisition of Arcelor, Mittal expected cost reductions to reach USD 1 billion within three years' time. Typically, the synergies targeted in steel company mergers are around 3% of costs.

Optimizing the Allocation of Production

Synergies are, at least in theory, relatively easy to achieve. However, whether management can properly identify and implement these synergies is another question. Even when synergies are not feasible, costs can be reduced by rationalizing production, that is, by shifting steel production from high-cost mills to more efficient mills following a merger, so long as the more efficient mills have excess capacity.

This type of efficiency gain is different from a synergy, since the merging partners' assets essentially continue to be used separately following the merger. This rationale for merging has been cited in numerous recent cases. For example, Tata Steel's offer to buy Corus was based, at least partly, on the cost-efficiencies of Tata providing slabs produced in India from captive iron ore at up to half the cost of UK-produced slab. In the Evraz–Oregon Steel Mills merger, costs could be lowered by Evraz supplying slabs produced in Russia at low cost using the company's own iron ore at Oregon's plate mill. This is envisaged to boost profit margins for Oregon's plate- and pipe-making operations. Moreover, ThyssenKrupp, Baosteel, and Dongkuk are involved in slab production in Brazil, while Posco and Mittal Steel have projects in India.

Steelmaking raw material prices have surged in recent years. For example, the price of iron ore has been up almost 400% compared to its level in 2000. Coal prices have also increased noticeably. As a result, many mergers and acquisitions have been driven by the desire to produce basic steel in low-cost regions near raw materials, yet maintaining or accessing geographical proximity to major consuming markets. A prime example was the recent bid for Corus by CSN and Tata Steel. CSN's rationale was that it could supply all of Corus' iron ore needs through its own mine in Brazil.

Greater Flexibility in Labor Contracts

Other cost benefits from acquisitions can result when the acquiring firm is able to lower labor costs by renegotiating more flexible contracts with the employees of the acquired firm. This has been the case particularly in the United States, following the wave of bankruptcies in 1998–2001, which forced unions to accept lower wage costs. In the case of International Steel Group's acquisitions of the LTV Corporation, the company negotiated a labor agreement with the United Steelworkers allowing for greater outsourcing activity and fewer job classifications, as well as a restructuring of compensation and pension plans. Allowing workers to perform a wider array of duties than before and for outsourcing during peak periods of demand ultimately boosted labor productivity and thus helped to reduce unit labor costs.

Attracting Capital

For a long time, capital markets were reluctant to commit resources to a steel industry suffering from chronically low profit rates, high costs, excess capacity, and at times bankruptcies. As a highly fragmented industry, the steel sector lacked the capital access to invest in new technology and in new products, to compete with alternative materials, to attract management and technical talent, and deliver what customers required when they required it. Thus, consolidation may be the means of permanently increasing profitability in the steel industry and help it attract capital for innovation and future growth.

Dynamic Efficiencies

Mergers in the steel industry could, in theory at least, give rise to so-called dynamic efficiencies. These relate to efficiencies that could be achieved through research and development or sharing knowledge and skills, which lead to the development of new products, production processes, or improved product quality and service. Consolidation may encourage steel companies to engage in more research and development activity, because there are fewer competitors to free ride on the benefits generated from their innovations.

The above prognosis by the OECD Steel Committee is plausible but probably a bit on the optimistic side.

Perspectives on the 21st Century Steel Industry

A new reorganization of the steel industry is now underway. While international steel trade and multinational steel company operations have been around for over a century, the industry is now organizing at the global level in a new way. Currently, steel companies are buying up coal and iron ore properties to reverticalize so that the tail does not wag the dog, that is, the monopolistic raw material producers do not hijack all the profit margins. Second, it is likely that steel producers will reacquire or reinstitute their own service center operations, principally so that they can get closer to the customers and increase the knowledge transfers that are critical to pushing out into the marketplace the new advanced steels that have been developed. These new service center operations, however, will not be a replay of the commodity broker role of the past but in fact become what academics call Knowledge Intensive Business Services (KIBS). To coordinate all of this activity, the reconstituted steel companies will be heavily dependent on IT and performance benchmarking as a new coordination platform in the industry. National steel industries as we have known them in their two previous iterations will be going away. The steel industry as a major contributor in national economies will not be going away. What we will have is national and regional configurations of steel facilities and capabilities operating within the new global steel supply chains.

Herrigel makes five points that differentiate his steel narrative from the conventional story of the American steel industry.

1. In the New Deal political economy of oligopolistic mass production, price setting was essentially a function of collective bargaining.
2. It was the minimills in combination with the service centers that put the integrated mills into the crisis situation with imports.
3. In responding to similar global pressures, in the USA, protectionism was the policy response of the integrated industry, different from the cartelization approaches in German and Japan.
4. The restructuring in the 1980s and 1990s was a negotiated restructuring, a non-market-based approach. Independent economic agents responding to price signals do not match the historical case.
5. The combined result of the process, in fact, has had the result of recasting the boundaries of the firm.

In the mid-1990s, at the height of the minimills-displace-integrateds enthusiasm, there was a very different story with different expectations for the outcome from the Herrigel account. It told a compelling story of the steel industry getting its act together in response to a dramatically changed economic and technological environment. However, the world of steel changed again in dramatic ways within 2 years. In 1997–1998, the Asia–Russian Financial Crisis tipped the scales in another direction for restructuring of the global steel industry. The sudden and huge drop in currencies took the floor out from under the world steel prices and undermined much of the reconstituting of the steel industry, which had taken place in the 1990s. Some of the most productive mills in the world, such as Korea, saw their costs drop by 60–70%. Steel prices fell and cheap imports surged, particularly as the US dollar rose significantly. What ensued was 5 years of dramatically lower steel prices, which had the effect by 2001–2003 of driving dozens of North American steel companies, both integrated and minimill producers into bankruptcy.

This was then followed, with the rise of China as a key market and huge consumer of raw materials inputs of the industry, into an unanticipated surge in steel prices from $250 to $750 a ton in North America from mid-2003 to the end of 2004. It also brought on the unprecedented wave of consolidation of steel companies, which we see unfolding to this day.

The period of steel restructuring discussed in the 1980–1995 period saw the global industry undergoing what the Europeans called a Manifest Crisis characterized by an overhang of world steel capacity of 100–200 million tons in excess of demand, declining prices below the cost of production for much of the period and steel priced entirely as a commodity. What a difference a decade makes. Taking into account the staggering growth in China and India, after chronic overcapacity since the 1980s, there was actually a shortage of steel capacity globally by 2005.

Current commentators on the steel industry speak of the reconstituting of global steel on regional lines.[5] The three major steel markets in the world—Asia, Europe, and North America—and the new global steel companies needing to have an active presence in all of them.

The steel industries of the BRIC in the first decade of the new century have been destabilizing the previous global steel industry configuration dominated by the USA, Germany, and Japan. China already destabilized the position of raw material inputs. Now, Brazil is shifting the boundary between raw materials and steel processing at the slab stage. India as it brings on 100 million tons of capacity in the next 10 years could trigger a resurgent export-led trade war.

We can expect that within this next decade, there will be a half a dozen truly global steel companies, each producing 100 M tons of annual capacity, and technological innovation within the three dominant steel regions—Asia, Europe, and North America—will be key.

The next 10–20 years of steel innovation will be led by integrated steel producers using iron ore and coal inputs and conducting a forced march of higher value-added steel products to satisfy new customer needs. The so-called dual-phase, high-alloy, high-strength steels will become widely applied in the market place. Some materials and products will even combine steel with other materials into new composite materials such carbon fibre. The new global steel industry will not be an island.

CHAPTER 5

Industry Organization and Competition

Industrial organization is the field of economics that builds on the theory of the firm in examining the structure and boundaries between firms and markets. In this chapter, we apply this lens to the steel industry and issues such as transaction costs, price competition, and barriers to entry. Three examples are used to illustrate the qualitative changes that have come to traditional steel producers: the EAF challenge, the Japanese Steel Revolution, and the impact of lean production on distribution and fabrication. In each case, the economics of the steel business has changed, primarily driven by industrial organization and social factors.

As outlined previously, classic Big Steel was built on oligopolistic competition and administered prices. It was famously called the Pittsburgh Price system. That began to fracture in the 1960s due to competitive pressures internally from minimills and externally from Japanese and European imports. Internally, the steel companies conceded low value-added market share to the minimills. Externally, they relied on litigation and Steel Trade Wars. But, in the past 30 years, innovation in steel became a critical driver of industrial organization and competition.

The trajectory of production capacity and employment of traditional steel mills is not a reliable proxy for the state of the steel industry as a whole. It is useful to remind ourselves of the product market segments served by different agents in the steel industry as illustrated in the following graphic.

2010 Steel Shipments by Market Classification

Source: American Iron and Steel Institute 2012.

At one time, the large producers served all of these markets. Much has changed. Declines in the primary mills have been matched in many cases with expansion in the service centers and the fabricators. Fundamentally it reflects an imbalance between the steel mills' batch modes of production and the order and inventory dynamics of the auto and other manufacturing industries and similarly in construction. The service centers in particular have essentially become inventory managers and suppliers of working capital to firms in the downstream industries.

It is estimated that more than half of all steel production globally is going into construction, especially when the latter includes infrastructure projects in emerging countries. In addition, construction is expected to be the largest growth market in the coming decade, even for advanced economies. But to take advantage of this, the steel industry is challenged to do things differently in terms of how steel is marketed, serviced, and sold. This is making fabrication the lead interface between the steel industry and the design and information systems of the new digital economy.

The result has been to produce different forms of industrial organization within the composite entity we call the steel industry. In turn, these entities compete for different product market segments.

Rise of New Competitors: The Minimills

The most fundamental competitive challenge to traditional steel producers arose from the minimills who posed not just a contest over price but also a completely different organizational model of a steel company and its operations.

Electric arc furnaces (EAFs) have been around since 1906. They displaced the inefficient, fuel-gorging crucible furnaces for specialty stainless steel, but whose annual US output averaged only about 120,000 tons of steel. Even at the outbreak of World War II, EAFs were producing less than 2 million tons annually. Open Hearth production had soared to more than 60 million annual tons in the same time frame. But the war demanded more steel, particularly alloy steel for armaments, and the quickest way to achieve this was to build EAF furnaces. In 1943, the US government sponsored the first Electric Furnace Conference. Several integrated steelmakers received subsidies to install EAF capacity and there was a temporary surge in EAF production.

In the postwar period EAF steelmaking remained the "poor relation" within the integrated steel plants, as open hearth output reached 86 million tons annually by 1950. However, the production of stainless and high-alloy steels remained with the EAFs. The tide in the industry however was about to change based almost entirely on the actions of two entrepreneurial individuals, Jerry Heffernan and Ken Iverson.[1]

A Canadian, Jerry Heffernan would emerge as the leader of the renewal of the EAF steel industry. He raised capital for the construction of Premier Steel in Alberta in 1954, where he wanted to implement his vision of displacing ingot casting with continuous casting. This was not a new idea. The technology was developed primarily for lower melting-point nonferrous metals like copper that were easier to process. Heffernan had a billet caster built by Rossi/Koppers at Premier in 1959 to meet the growing demand for sucker rods for the oilfields. But most of the steel was still cast into ingots. Meanwhile, Stelco, the U.S. Steel equivalent in Canada, saw a Western market slipping away, and eventually bought Premier from Heffernan in 1963.[2] By leveraging this capital, he was able to construct the Lake Ontario Steel Co. (LASCO) in 1964, near Toronto. If a minimill is defined as one where 100% of the output from an EAF shop is continuously cast with no ingot mold backup, then LASCO was the first minimill not only in North America, but also in the world. Other plants had casters, but none was as completely committed to the continuous casting process.

Although LASCO was a union mill, Heffernan was able to introduce a new management philosophy into steelmaking—a limited supervisory hierarchy with emphasis on continuous improvement in all activities. This encompassed not only process improvements, but also training, better safety, and incentives to involve the work force in expanded responsibilities on the shop floor. It was McGregor's 1960 "Theory Y" of management put into practice, another first for the North American steel industry. From this beginning, the new minimill culture has challenged the traditional North American steel industry and arguably has had even greater impact on productivity than all but a handful of technical innovations.

To jump ahead in the story and connect the dots. Heffernan became the mentor to a young Ken Iverson, who would go on to turn the American steel industry upside down.

With Lasco running well, Heffernan now built a minimill in partnership with Cargill in St. Paul, MN, known as North Star Steel. At the same time, Ken Iverson, after a series of positions in diverse metalworking and casting companies, became almost by default the president of the struggling Nuclear Corporation, whose only profitable operating unit was the Vulcraft division in Norfolk, Nebraska. When his sole supplier of bar steel for this fabrication plant kept raising prices, Iverson decided to seek another source of rebar. The North Star mill was the closest bar mill to Norfolk, and after contact with Jerry Heffernan, Iverson decided to build his own mill to supply his own steel. The teetering Nuclear Corp. was now headquartered in Charlotte, but Darlington, S.C., was selected as the site for its first steel mill because it was close to a Vulcraft joist plant. The mill was commissioned in 1969 and was designed for a modest annual production of 200,000 tons. Learning how to continuously cast steel at Darlington with a "green" crew of farmhands nearly put Nuclear into bankruptcy, but by 1971, experience had been acquired, the bugs had been worked out and steel sales in 1972 turned Darlington into a "gold mine." The company name was changed to Nucor. Two more mills at Norfolk (1974) and Jewett, Texas (1975), piggybacked this success.

All these mills were also built at rural sites, where hardworking farm workers with good mechanical skills and no preconceived ideas about steelmaking were eager to work for a low wage but big bonuses based on production volumes.

By 1975, the 30+ minimills in the United States and eight more in Canada spanned the continent and produced about 6 million tons annually of bar and section products. From the perspective of Big Steel, they were not particularly threatening—long product markets were being eroded, but the flat rolled markets were clearly safe. The integrated mills had other problems to worry about.

Changes in the regulatory environment started to shift the incentives and capital allocation decisions in the industry. A series of legislative measures in the early 1970s had huge impacts on the steel industry: the Clean Air Act (1970), Clean Water Act (1972) followed on the heels of Occupational Safety and Health Act (OSHA) and were to be followed again by new pension regulations in the Employee Retirement Income Security Act (ERISA). These mandated heavy capital expenditures and increased operating costs to meet environmental and safety regulations, and impacted the integrated mills to a far greater extent than the minimills. Imports were rising, unions were militant, pensions had to be funded, and while 1973 was a record year for raw steel production in the United States, it was also the year of the first oil crisis. The delayed economic impact of this was felt in 1975, as shipments dropped by 20%, with the brunt of the reduction taken by the integrated mills. It was a body blow from which Big Steel never fully recovered.

Minimill Breakthrough: 1989–2005

The lean management hierarchy in minimills accelerated the evaluation and implementation of foreign as well as home grown ideas. It was no secret that the Siemens group in Germany had been operating, for several years, a pilot-scale thin-slab casting operation, where slabs of steel are produced directly from the furnace without having to be rolled from ingots. The facility had been visited by numerous groups from the United States, including the major steel companies. But Iverson of Nucor had the nerve to abandon a $5 million previous investment and install the

Siemens thin-slab caster in the cornfields of Indiana at Crawfordsville. The year was 1989, and the commissioning of this compact strip process (CSP) qualifies as a turning point in the history of the steel industry, ranking with Bessemer's process in 1856.

In the CSP, a single-strand caster with a special mold produced a continuous 52-inch-wide strand of 2-inch-thick steel that is sheared, fed into a tunnel furnace for temperature control then passed directly into the rolling mill. This came very close to continuous steelmaking.

The Nucor Blytheville structural mill—a joint venture of Nucor with Yamato Steel, called Nucor-Yamato Steel, that was commissioned in 1988—was running exceptionally well, and Crawfordsville amazingly turned a profit within a year. Nucor built a second and even more productive mill at Hickman before anyone else in the United States had even one thin slab caster.

With the precedent set, "greenfield" sites producing sheet, plate and structural products were built all over the United States. Between 25 and 30 million tons of capacity for these products alone were added since 1990, to offset the millions of tons of integrated capacity that was closing down. These were multimillion ton mills and thus rendering the term "minimill" somewhat obsolete.

Energy requirements and greenhouse gas emissions per ton from EAFs were the lowest in the world, while manpower was reduced in some cases to the unheard-of level of less than 1 MHPT.

The newer Nucor Steel–Berkeley facility installed facilities for the production of ultralow-carbon steels. It meant that tinplate and special electrical steels remain the only untouched flat rolled products not targeted by the new EAF plants, which already had come to dominate plate, structural, and rail production in the United States.

In summary, the Nucor revolution produced a new form of industrial organization in steel to penetrate new product markets formerly controlled by the traditional producers.

The EAF picture is not trouble free. Prime scrap availability and pricing are two potential concerns for the minimills. There have been sharp and unanticipated jumps in the average price of premium quality scrap. Spikes for prime scrap on a monthly basis have been as high as $400/ton in recent years.

The Japanese Challenge to Industrial Organization in Steel

The Japanese threat to American steel producers was as much an issue of industrial organization as it was price competition.

The best analogy is to Henry Ford's assembly line revolution at the Highland Park plant in 1913–1914 where he reduced the price of the Model T Ford from $900 to $300. Forty years later, the steel industry equivalent took place along Tokyo Bay with the Kawasaki's Chiba steel mill in 1952. There, notwithstanding the previous development and language of "integrated" steel mills, Kawasaki built an entirely new kind of steel plant that was physically organized and internally coordinated in a revolutionary new way. The paradigm shift was reflected in the internal railway system. Where a typical world class mill had over 150 miles of internal rail lines, the Chiba plant had less than 60 miles. This fundamentally changed the flow of raw material inputs and steel processing. It meant a shift from batch to continuous flow production that would lead to a new world of steel production and steel products. It became the metallurgy and production system to support the Japanese quality revolution in manufacturing that became visible to consumers a decade later.[3]

The Chiba experiment in its first stage, began with the old furnace technology, the Open Hearth. These were later replaced by the newer BOFs. The difference in Chiba was in the overall layout and integration of the whole steelmaking and processing flow. With this revised architecture of steel mills in hand, the revolution moved to the 'hot end'—the hot metal producing part of the mill. The productivity of Japanese mills quickly became multiples of what their European or North American competitors were capable of. The change began with the building of new high capacity, large diameter, Blast Furnaces but it quickly spread to the rest of the steel mill. The traditional multistage process was costly and energy intensive. Achieving and maintaining quality was a constant challenge.

The continuous casting machine (CCM) became the means to simplify and more closely integrate many of these steel production steps. The product from the BOF was poured directly into a mold that produced a constant stream of slabs which could then immediately be moved to the finishing stages for transformation into plate or sheet. The BOF-CCM configuration reduced direct costs by 30–50%, produced a continuous stream of product and opened new avenues for controlling and improving quality. Steel production flows began to resemble a chemical plant more than discrete manufacturing stages.

The new continuous steel production process, in addition to eliminating dozens of individual steps in the process, also unleashed such volumes of steel that the flow could not be dealt with by the traditional mechanical and human processes. The new, higher volume steel throughput drove the next technology step which was the introduction of computers to the steelmaking and processing stages. The first large scale introduction of computers for production purposes began in the late 1950s and early 1960s, again in Japan. Another steel mill reference site emerged as the global benchmark, the NKK plant at Kimitsu, just around Tokyo Bay from Kawasaki Chiba.

The combination of computers with improved control systems now gave steelmakers the physical means and the data to continually monitor metallurgical qualities of the steelmaking and finishing in real time. This was the second stage of the Japanese steel revolution. It would become the keystone of the materials infrastructure underlying the quality revolution in the Japanese auto industry and ultimately in global manufacturing that we all now take for granted.

The quality revolution in steel was not simply a matter of machines and metal. The human element and social organization soon came under the same challenges as traditional technical steelmaking. The traditional batch production steps of making steel in the Open Hearth, monitoring it and getting ready to tap it for the moulds, etc., had as its complement a very intricate hierarchy of skills, occupations and social statuses. The oversight at the face of the furnace was controlled by the Lead Hand, a highly skilled worker with a team of Second Helpers, Third Helpers, Laborers, etc., along with the inevitable Foreman. At regular intervals they would peer into the furnace and judging by the color and texture of the flame, would decide on adding different fluxes and charge (limestone, scrap etc.) to achieve the desired type and quality of steel. They would then take samples of the molten steel and pass these to the laboratory. When the steel met the technical specifications it was ready for the next processing stage. The steel wouldn't be released until the engineers and the technicians in the lab judged it to be fit. This time honored system of procedures and skills hierarchies was applied around the world.

However, as the Japanese mills scaled up they found that this whole system of procedures and work organization simply couldn't keep up with the increased flow and pace of BOF steelmaking. New instrumentation and continuous monitoring, enhanced by computers, came to replace the whole traditional organization of the shop floor.

Production workers and the metal itself couldn't wait for the engineers, lab technicians, etc. Responsibility for production control and ultimately quality control started to pass from the engineers to shop floor workers. And, given the connection between quality and products, this development was soon seen as strategic for management and the company as a whole. As a result Kimitsu soon became the first site for the development of work teams and quality circles in the steel industry.

Kimitsu was a tipping point. It embodied the fact that the information economy was emerging in steel mills in the 1960s and 1970s, 20 years before it became a common term for society as a whole. As volumes increased and downstream technology developed, computerization and shop floor skills evolved and the Japanese steel industry became the reference point for best work practices around the world. A whole new perspective developed about the steel mill and technical innovation. The approach that dominated the first 75 years of 20th century steelmaking— that technical innovation would take place in specialized industrial laboratories of the German-Siemens model and then transferred to production facilities for implementation—was challenged. The production plant itself came to be seen as the site of, or at a minimum, a co-developer of new technologies.

None of the Japanese steel companies had ivory tower research laboratories or R&D sites. They all had their labs in close proximity to or literally inside their production plants. Interestingly, Dofasco which would become the most profitable North American steel producer, always kept its lab in the plant. However, this changed perspective on steel innovation involved more than just labs and production plants. The success of the Postwar Japanese steel industry was not simply a function of individual engineers, managers, and companies. The newer steel production innovations such as continuous annealing originated with teams of production workers on the shop floor and not with engineers and materials scientists (Vincenti 1990).[4]

Steel Distribution and Fabrication

The third example of industrial organization change in steel came downstream with the associated subindustries in the steel supply chain.

The Broader Steel Sector comprises the Primary Steel Producers (NAICS 3311), together with mills that produce pipe and tube or roll and draw iron and steel (NAICS 3312),[1] and three industries that historically have been linked closely to Primary Steel Producers:

- Iron and Steel Foundries (NAICS 33151),
- Construction Fabricators (subcomponents of NAICS 3323) and,
- Metals Service Centres (NAICS 4162).

Metals Service Centres inventory and distribute semiprocessed steel products to downstream users of these products in the manufacturing and construction industries. Over the past two decades many Metals Service Centres have also taken on 'finishing functions' which were previously carried out by the Primary Steel Producers. Examples are: sawing, shearing or cutting basic steel shapes into standard sizes, rolling basic steel shapes to produce angle products, drilling, threading, slotting, and painting. Metals Service Centres obtain steel from domestic producers, other NAFTA region producers and from off-shore sources. For many steel-using companies, their primary contact with the steel industry is through Metals Service Centres who also supply nonferrous metal products.

---

1. Some pipe mills, rolling mills, and drawing mills are owned and operated by Primary Steel Producers. Other mills are independently operated. Mills that are operated by Primary Steel Producers are included in NAICS 3311. Mills that are operated independently are included in NAICS 3312.

Foundries re-melt steel, cast the molten steel into specific shapes such as wheels for railway cars and automotive engine blocks and further finish the cast product by grinding, sanding, drilling, etc. Historically, many foundry operations were owned by Primary Steel Producers, though this is no longer the case in North America. While some foundries work solely with iron and steel, it is increasingly common for foundries to cast other metals, especially aluminium. Some foundries make their own molds for casting, while others purchase molds from specialized mold-making shops. Over the past decade, the majority of ferrous foundry operations in North America have closed or moved offshore.

Construction Fabricators comprise plants that fabricate plate work and structural products by cutting, punching, bending, shaping, and welding steel for use in the construction industry and in other heavy industries, notably the mining and energy sectors. Some companies that are considered Construction Fabricators also manufacture architectural iron and steel products, such as staircases. Construction fabricators are the principal channel through which structural steel products enter the construction market.

For the whole Steel Sector, a fundamentally new challenge has arisen in the change in the rhythms of steel-consuming industries. Steel mills, in their iron- and steelmaking departments, are still fundamentally batch-mode facilities, with an underlying imbalance in the production and scheduling cycle times with their two major consuming industries—manufacturing and construction. Delivery times for the 'sweet spot' in mill rolling schedules is 4–6 weeks. The best claim to do it in 30 days. Meanwhile delivery time in auto production schedules is 3–4 days and in construction fabrication it is 1–2 days.

This new time factor became a critical factor in industrial organization by re-distributing both functions across the industry. In addition to the normal impact of productivity gains over time, most of the "job loss" associated with primary steel mills has actually been a transfer of labor and functions to auxiliary industries because of the time factor.

Traditional processing functions like cutting coils and sheet into smaller "blanks" for manufacturers shifted to service centers. Distribution functions also moved away from the primary mills. Therefore much of the "loss" of the steel industry has been a redistribution within the larger steel

Employment in the Broader Steel Sector, 1991–2010[6]
Statistics Canada, CANSIM 281-0024

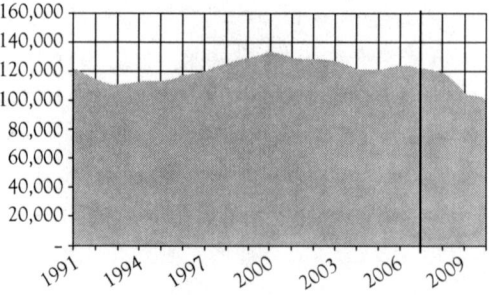

Monthly Employment in the Canadian Steel Sector
January 2006 to October 2010
Statistics Canada, CANSIM

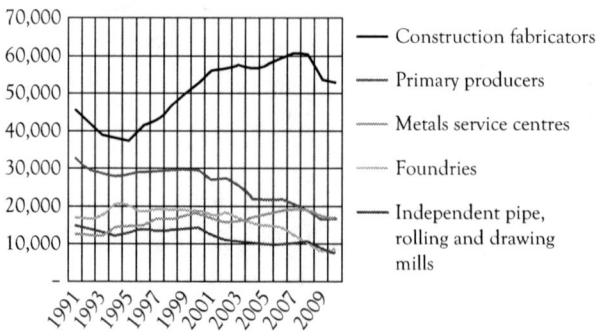

Source: O'Grady & Warrian (2011).

cluster. It also called forth different business models in different parts of the steel supply chain.

Taken together, all these factors of shifting industrial organization have had an equal or greater impact on the steel industry and its companies that have the customary drivers of change—technology and trade.

CHAPTER 6

Steel Labor Relations

The steel industry has a long and difficult history of labor relations issues. The systemic inability of the two parties to resolve their differences at the bargaining table and the resulting series of strikes in the 1950s and 1960s, in particular, is now seen to be the tragic failure of the American integrated steel industry, opening the door to minimills and importers to undermine product and pricing mechanisms that constituted ground zero for Big Steel. For that reason, this chapter gives extended coverage to the mechanisms and issues that lay at the heart of the conflict.

Perspective on Steel Industry Labor–Management Relations

As mentioned above, the steel industry has been one of the icons of the mass production industrial age and industrial unionism. Huge facilities and huge investments in capital equipment have been the characteristics of the business and at times the bane of its existence. Skills within steel companies were basically organized around a tight hierarchy of engineers and executives at the top. Very few CEOs of steel companies in the 20th century did not come from the engineering staff. Beneath this hierarchy, there was a mass of unskilled and semiskilled industrial workers and laborers.

Frederick Taylor, who wrote the famous *Principles of Scientific Management* in 1911, did his original work in the steel industry and sought the complete decomposition of jobs into simpler, less skilled components. The steel industry took this philosophy to heart and tried to implement it more systematically than any other major industry. The eventual codification of the whole system of jobs in the steel industry, the Co-Operative Wage Study (CWS) system, gave skills and wages in steel a uniquely hierarchical and fragmented character that still largely defines the industry today.

On the other side, the cycles of collective bargaining in the steel industry include a number of false starts within a long-range objective on the union side to move away from its classic adversarial role and gain the right of the union to a voice in the industry. The USWA leadership thought that they were sold a bill of goods by the companies in the steel crisis bargaining of the mid-1980s. Though later they largely made those concessions whole, the parties reentered the change cycle in the early 1990s when ongoing difficulties in the industry compelled a new, different, and union-led agenda for collective bargaining. It was partially successful. The early adoption of these new directions took place in the smaller sites installed as part of the US–Japanese co-ventures but there was very little change in core steelmaking operations of the major steel companies. It was only after the 2001 crisis and a new cycle of bankruptcies that the opportunity for more union voice arrived, with additional contract changes and roles being forced on to both the labor and management groups.

The Labor Process in Steel

The steel industry in the nonunion era was dominated by the impact of technological change on workers and the production process as the industry entered the mass production age: The mechanization of steel production process in the early 20th century altered skill requirements in important but often unexpected ways. Machines replaced human muscle power in moving materials through the production process and thus speeding up the process. But, on the whole, instead of breaking down old crafts, machines had simply replaced some older, heavier manual routines. Blast furnaces were much larger, but still involved fundamentally the same chemical process. Open hearth furnaces had surpassed the limited output of puddling furnaces, but not the need for metallurgical knowledge. Rolling mills of massive proportions could handle much more metal, but still needed the careful attention of experienced workers. As a result, the old dichotomy of craftsmen and laborers gave way not to a mass of mindless machine-tending jobs, but to a new hierarchy of jobs with less of a gap between the least and the most skilled and with considerable demands for responsibility and judgment from workers filling them, perhaps with greater occupational homogeneity.

Previous analysis of these developments that emphasizes simply the stripping away of shop floor discretion misses the point.[1] Close-knit, co-operative work teams at the center of the production process headed by a few key workers like blast furnace keepers, open hearth melters, and head rollers exercised a range of independent decision making on the job and a freedom from either direct supervision or machine-pacing. They were more similar to the shop floor status of the 19th-century craftsmen. For many more, the range of their discretion was much more narrowly circumscribed but nonetheless crucial to the most efficient operation of the machinery. Overall, given the disappearance of so much unskilled labor, the average skill level in the new steel plants probably went up, not down.[2] Operators and rollers saw themselves as unique and more closely identified with their Mill or machinery than with the Company. This would present a large hurdle to be overcome when Statistical Process Control (SPC) and quality measures were introduced in the 1980s and 1990s.

In the first decades of this century, gone were most of the shovels, wheelbarrows, and small army of brawny laborers and in their place were a few handfuls of men who manipulated gears. Under the old style it took 150 men per 24 h to operate a 200-ton furnace and the output was 1.33 tons per man. Under the new system, it took only 60 men per 24 h to operate a 550-ton furnace and the output was 9.17 tons per man.

The companies, however, had done more than simply change individual work processes. They brought them together in huge sprawling complexes that tightly linked each phase of production with the others. Individual departments and processes within departments nonetheless retained some distinctive rhythms and occupational requirements. Some processes were naturally linked, notably the coke ovens, blast furnaces, and open hearths, while others especially the smaller rolling mills were based on batch production. Some, like blast furnaces and rolling mills required steady, routinized feeding of furnaces or machinery, while others required more erratic bursts of frantic exertion, especially tapping furnaces. There was, in short, a great variety of work processes in a steel plant that did not necessarily lead to a common occupational experience for workers in the industry. There was an informal pecking order in the mill and Steelmaking was King.

In many ways, the sum of all the technological innovations was greater than its parts. Besides the changes in individual stages of production, it was the overall coordination of the plants that was so remarkable. Every plant was a maze of tracks for numerous railways, cranes, and conveyors. Raw materials moved along these tracks to coke ovens and blast furnaces; liquid pig iron was swept off to the open hearths at the end of giant cranes; steel ingots were shunted off to the rolling mills, where cranes and conveyors carried steel forward. In contrast to past practices, there was far less remelting and reheating of the metal as it moved through the plants. Early 20th-century steelmaking did not involve an assembly line, but there was definitely an integrated flow-through. For the most part, the plants ran continuously through the year, rather than on the seasonal basis that had characterized much of 19th-century production. In fact, several departments ran nonstop 24 h a day and 7 days a week. Mechanization therefore brought not only greater volume of production from the new facilities, but also greater speed and intensity and, for the workers, greater pressure to keep up.

The Steel Industry Wage Structure

People with only a little knowledge of steel industry labor contracts are stunned by the detailed and complex wage and job structures they observe. Few need more convincing that "rigid" labor contracts are a major problem in the industry than to read the wage clause of the collective agreement. In the steel industry, this whole system is known as the Cooperative Wage Study or CWS as it is called. It is now regarded by most as a union construct; however, in fact, the wage and job structures originated with the company management prior to being incorporated into collective agreements.

The origins of CWS go back to a World War II US labor law decision to address wage disparity disputes. Contrary to widespread belief, the Labor Board did not order the parties to develop a joint job evaluation program. In fact, neither the board nor the steel panel expected that the job evaluation route would be used by the parties. There is even less in the record to indicate that systematic job evaluation was a practical possibility in view of the union's hostile attitude to any such approach.

The Cooperative Wage Study was set up in Pittsburgh by a group of 12 steel companies in 1943. It was exclusively comprised of company representatives, assisted by private consultants, the American Associated Consultants Incorporated. The stated objective was to "determine the wage-rate situation in the companies; determine what it should be; and determine ways and means by which to bring about such corrections as were found to be necessary."[3] Again, contrary to widespread belief later, the union was not a part of the original design of CWS; in fact, it was generally opposed to such undertakings as a preemption of collective bargaining.

The basic objective of CWS was to discover the underlying wage structure of the steel industry—not of any particular plant or company, but of the industry as a whole—and to construct a manual that would reflect this structure and identify deviations from it. These variances would be considered as inequities from the underlying wage structure, to be marked out for elimination through processes agreed upon in collective bargaining with the union. From the outset, the CWS approach was predicated on the assumption that there existed a general wage structure for the basic steel industry, which, in varying degrees, would reflect job relationships in the individual steel plants and companies. The individual plant and company rates for a large number of jobs, when assembled, supported the thesis that there was a general wage structure for the industry and indicated that the companies were on the right track in working toward an overall industry program rather than individual company plans to eliminate wage inequities. In contrast to the usual approach to job evaluation, which tries to "correct" a wage structure through the application of a manual with preconceived weights, CWS architects believed that designing a wage structure was beyond the scope of job evaluation. They preferred to find and use the weights for various job factors that had been developed in the steel industry through the impact of the common labor market, the ups and downs of business cycles, hundreds of thousands of individual judgments and individual bargaining as well as collective bargaining.[4]

Factor weighting is considered to be a trial and error process, with a given set of weights subject to testing with benchmark jobs. Adaptation of existing weights to a particular configuration of benchmark jobs or even modification of the maximum points allotted to factors is not uncommon and is considered good job evaluation practice. Once the factors and weights have been determined and the benchmark jobs evaluated, the remaining jobs are interpolated by reference to them.

As a matter of preference, it is not necessary to use given factor weights at all in job evaluation. In the factor-comparison method of job evaluation, no fixed weights are assigned to individual characteristics of a job. Factor weights are developed from the benchmark or key jobs so that the accepted total rate for each such job is distributed among the factors in terms of cents per hour rather than points. A feature of this plan is that there need not be a range or a limit set for any job characteristic or factor and predetermined degrees need not be defined. When the factors have been assigned their money weights for the given job, it is necessary only to add the amounts for all factors and the sum becomes the money rate for the job. An advantage of the factor-comparison method for an industry like steel, where there is an unusually wide spread in rates, is that it allows unlimited room at the top for any exceptional factor. Jobs are ranked in each factor by comparing their relative worth to that of the key jobs. This is a flexible method that permits significant differences in the relative weights of factors for different applications of the plan. Thus, the CWS approach was not unique in its refusal to accept fixed factor weights in the evaluation of steel jobs. It was unique, however, in the method it used to arrive at relative factor weights that would best reflect the steel industry wage structure.

Most other job evaluation plans had been constructed around the wage structure typically found in light manufacturing industry. Such plans typically allocated 50% or more of the maximum points available to skill factors and relatively little weight to responsibility, effort, and working conditions. This distribution of factor weights in other industries resulted in highly skilled jobs such as tool and die maker, pattern maker, and machinist being ranked at the top of the job hierarchy, which is generally in accord with their position in the plant wage structure across general manufacturing.

Steelworkers in the mills however had a great deal of responsibility for materials. Executives themselves said that 90% of the operational problems in steel mills involved the flow of materials. Workers also still required major physical exertion and labored in unusually difficult conditions of heat, dust, and danger.

However, the steel industry was also characterized by highly interdependent processes performed in a group rather than on an individual basis and involving the use of heavy and expensive equipment. Therefore there are a number of jobs that rank higher than skilled craft occupations in the steel plant age hierarchy. The main difference between such steel jobs as Roller—Bloom Mill; First Helper—Open Hearth; Blower—Bessemer and Heater—Plate Mill, as compared with typical skilled maintenance jobs, is that in addition to a high degree of skill these jobs also involve considerable responsibility for the continuous flow of operations and for expensive tools and equipment, which is not normally required of tradesmen. Under conventional job evaluation plans, such production jobs will hit the ceiling provided by the maximum points for responsibility factors without attaining any significant differential about highly skilled maintenance jobs. These jobs would be adjusted out of line on the high side and their rates red-circled under most plans, despite the fact that there is general agreement among the union, management, and the worker themselves that top skill production jobs deserve to be ranked well above skilled trades jobs in the plant. The suspicion that radically different factor weights were needed to reflect the steel industry wage structure led to the search for an entirely new manual rather than an attempt to modify existing plans to fit the needs of the industry.

As stated, a distinctive factor in the CWS manual is "Responsibility for Operations." A steel plant is typified by a highly interdependent series of processes, most of which are performed by crews working with large machines and equipment. The smooth flow of work from one process to another is, therefore, of the greatest importance, because faulty operation on one piece of equipment has serious effects on processes both before and after that particular operation. This can lead to underutilization of both equipment and labor. This responsibility is apt to center on the top working member of the crew whose skill, knowledge, judgment, and motivation is typically the effective determinant of productivity. Roller, Heater—Soaking Pit, and First Helper—Open Hearth are illustrative. To meet this element, "Responsibility for Operations" was introduced. Later, as a result of the introduction of a new level in this factor during negotiations with the union, all crew members of production units having a share in the responsibility for operations received some credit under this factor. Below is an example of a CWS Job Description and its Classification from the 1950s.

CWS Job Description	
Department: Bloom, Billet and Rail Mills	Standard Title: Bloom Shearman
Sub Division: 44″ Mill—Rolling	Plant Title: Shearman
Date: December 22, 1952	

Primary Function

To operate a shear and direct the cutting of blooms, billets, and slabs to obtain best yields consistent with quality and customer requirements.

Tools and Equipment

12000 Ton, one-cylinder up-cut hydraulic shear and electric controllers for approach and shear tables. Hydraulic levers for air cushion, central guides, shear gap; wrench, hammer, bars, calipers, rule, etc.

Materials

Steel slabs, billets, and blooms.

Source of Supervision

Foreman. Not closely supervised.

Direction Exercised

Directs Shear Leverman as required.

Working Procedure

Receives specifications and schedule of items to be rolled from Shear Recorder.
Brings steel to shears on approach tables.
Adjusts gap between shear blades according to size of product.
Crop ends of blooms, billets, or slabs and cuts to desired length.
Measures section with calipers and reports to 25" Mill, Rail Mill, or 44" Mill Rollers.
Cuts out imperfections in blooms, billets, or slabs if necessary.
Cuts tests as required.
Uses bar to release any crop ends that jam in the chute.
Operates air cushion.
Operates wing guides to center steel on blade.
Checks blades and keeps blades secure.
Changes blades and assists on roll changes.
Performs work requiring a working knowledge of tolerances, heats, identification, and good yield practice.
Works in close co-operation with Deseamer crew.
Keeps working place clean.

Job Classification			
Plant Title: Shearman Standard Title: Bloom Shearman			
Factor	**Reason for Classification**	**Code**	**Classification**
1.	Preemployment Training. This job requires the mentality to learn to: Operate shears for a variety of sizes and use judgment in estimating amount of scrap to crop off.	B	.3
2.	Employment training and experience. This job requires experience on this and related work of: From 7 to 12 months of continuous progress to become proficient.	C	.8
3.	Mental Skill Use considerable judgment to operate shear to obtain maximum yield.	D	2.2
4.	Manual Skill Change and adjust shear blades. Assist on roll change. Controls movement of shear and passage of material to and from shears.	B	.5
5.	Responsibility for Material Estimated cost Use close attention while cropping slabs or blooms. Under $1000 Excessive scrap from too heavy cropping	C	2.5
6.	Responsibility for Tools and Equipment Exercise moderate care to prevent damage to shear blades and auxiliary equipment by cutting cold steel, spawling from lack of water or damage to entry guides.	Md. C	.7
7.	Responsibility for Operations Operate an Important part of a major producing unit.	E	3.0
8.	Responsibility for Safety of Others Ordinary care and attention. Directs shear blade changes. May operate shears where others are exposed.	B	.4
9.	Mental Effort Moderate mental or visual application required to operate shear, table controls, and determine number of cuts.	C	1.0
10.	Physical Effort Light physical exertion required to operate shear and table controls. Change shear blades and assist on roll change.	B	.3

(Continued)

(*Continued*)

11.	Surroundings	B	.4
	Inside—works near hot steel most of the time protected by shield.		
12.	Hazard	A	Base
	Accident hazard low. Exposed to some hazard while assisting on roll changes, blade changes, barring butts, etc.		
	Job Class 12 Date Sept. 17/54	Total	12.1

In this ranking scale, in a normal manufacturing plant laborer and machinist jobs sit at opposite ends of the scale. These jobs are found frequently in both light and heavy industries and should be evaluated properly by any plan. Jobs with exceptionally high skills, like tool and die makers, are favored by the light industry scale, while heavy industry jobs with high responsibility do not get proportional credit for the responsibility factors The net result is that jobs like Roller and First Helper are rated about the same or even lower than highly skilled craft jobs by light industry plans even though they have always earned far more than those jobs in the steel industry. In steel, the high end production jobs went to the top of the hierarchy.

The 1959 Steel Strike

The critical moment for steel labor–management relations was the 1959 round of negotiations, which saw the most pivotal of many strikes. The core was a dispute about the authority of management and work rules on the shop floor.

The 116-day strike by the USWA against U.S. Steel and the other large steel companies was the most important strike in the history of the industry. It has been seen as the all-time great strike over "restrictive work rules." Observers on all sides characterized the conflict as the classic industrial relations struggle over workers versus managements rights and authority on the shop floor. The fundamental issue was article 2-B of the collective agreement concerning local practices. However, the most comprehensive study of the 2-B provision reveals an ironic twist to the otherwise classic positioning of labor and management over employment security.[5]

Section 2-B of the agreement was first negotiated in the 1947 contract. At the time it received little attention from both the union and the employers. They were both more concerned about postwar wages and capacity utilization in recessionary times. However, by 1959, it had become the core issue in the struggle over shop floor rights and management prerogatives.

Section 2-B was a six-paragraph clause in the contract recognizing the existence of "local working conditions" either "practices or customs," "written or oral." It protected these local work practices and provided only two circumstances in which they could be changed: by mutual agreement between management and the union; or, unilaterally by management if the "basis" for working conditions changed. Subsequent arbitration cases made the explicit formulation that the meaning of "basis" for change had to be technological change or formal changes in work methods. For instance, arbitrators ruled that crew sizes were covered under Section 2-B. Management could not defend crew reductions solely by citing its management rights clause, work load levels, and time studies. Cost reductions alone could not justify a crew reduction. Management was however given permission to eliminate a job as long as it could prove that the basis for the job had changed, either by the introduction of new technology or by an alteration in the production process. The union frequently grieved in order to document a paper trail of working conditions and staffing levels. Gradually it created a written record of what staffing levels were at each level of operation at every plant across the whole industry, which could be invoked as a precedent later. The union argued that management had no right to eliminate or change a job solely on the basis of economic factors where no changes in technology or the production process were involved.

In fact, management across the steel industry in the 1950s had been eliminating thousands of jobs—employment in the industry peaked in 1953—by combining job functions or doubling up jobs, and not through technological innovation. The union believed management only wanted to slash jobs and speed up production, and management believed that the union wanted to interfere with all efforts at improving productivity.[6]

Section 2-B neither protected steelworker jobs from automation, nor limited management's ability to introduce new technology and machinery that would result in a reduction of staff or a reorganization of the production process. Section 2-B did, however, limit management's right to cut staffing levels and reorganize jobs without a corresponding introduction of new equipment or technology. The latter was precisely what companies like U.S. Steel sought to do in their older mills in the 1950s and why Section 2-B assumed such importance.

Having rejected the new technologies of the BOF and continuous casting, during the 1950s the steel companies were seeking to modernize by "rounding out" their existing facilities. In many cases, new equipment could not be introduced without seriously disrupting the product flow. Therefore, steel management went on an active campaign to reduce costs by reducing existing staffing levels and adding new responsibilities to jobs without introducing new equipment or technology. Section 2-B prohibited this and the union used the clause to block job slashing and production speed-up. As a result, management concerned about their increasingly inefficient and aging mills came to view the section as the disastrous unintended result of the 1947 bargaining, which not only limited management's authority to determine such critical shop floor issues as crew sizes and force reductions, but also undermined its cost reduction strategies.[7]

U.S. Steel negotiators tried in the 1959 bargaining to get a provision in the industry-wide contract to overrule local practices and informal agreements. Older mills had thousands of work practices. A national contract could not possibly codify all the individual work practices and informal agreements at all these mills. Even at the level of the individual mill, all local working conditions could not be summarized in an agreement, because local work practices and particular understandings were too numerous.[8]

The USWA leadership turned the strike into a crusade against a return to the conditions of the 1930s—complete unilateral authority of management on the shop floor and the end of job security. Eventually the union triumphed. They received a modest wage increase and the iconic Section 2-B remained.

With hindsight, there are two incredible ironies in the story of the 1959 Strike and Section 2-B. First, if management had indeed embarked on a major technology innovation program and pursued the BOF–Concast route they would not have been inhibited by the union contract. Failure to innovate would eventually lead to the companies' demise. Second, looking backward, the post-1993 contracts now explicitly require management to engage with the union and the workers over technological change and changes to production methods, the exact things that management rights in the mass production era were attempting to block.

The 1959 Strike had another political legacy. The next time steel bargaining came around in 1962, the parties again came to an impasse. John Kennedy intervened to avoid another major economic disruption. He convinced the union to accept a zero wage increase. He asked the companies in return to constrain prices. They initially rejected his demand outright. Then, under threat of a government takeover, the steel companies reluctantly agreed and a settlement was reached. However, the bitter taste and deep distrust between the steel industry and government persisted for years to come.

Centralized Bargaining in Steel

Centralized bargaining has always been the gold standard in trade union circles. The principal objective of centralized collective bargaining was to "take wages out of competition." Wage increases are more readily and securely achieved, and companies are forced to compete on grounds of efficiency and innovation, not on downward wage pressure.

The postwar steel industry was an outstanding period of high Wagnerism, named after Senator Wagner of New York, who introduced the New Deal labor legislation in the US Senate in 1935. The labor and management parties were to pursue their competing self-interests by negotiating complex collective agreements with multitudinous work rules. Both sides wanted it that way. Distributive bargaining and compliance-based employment relations were the norms of the day.

The steelworkers union was also strongly positioned for industry-wide bargaining because of the national-level jurisdiction of the federal National Labor Relations Board.

Mangum and McNabb[9] analyze the dynamics of the rise and fall of centralized bargaining in the U.S. steel industry in terms of the relationship of product markets and labor markets. According to the authors, the turn of the century steel industry was different from other oligopolies such as automobile manufacturing, in that steel had a higher degree of product homogeneity. Steel was steel and consumers were unconcerned with the source as long as it was cheap and in plentiful supply. There was also a substantial degree of technological homogeneity. Steel was produced throughout the industrialized world by almost identical coke ovens, blast furnaces, open hearth furnaces, and finishing mills, all staffed by workers with an almost identical complement of skills. In addition, because of its weight and consequent transportation costs, the product for steel was regional.

The skills required were virtually identical in every mill but the companies did not compete for workers within the same labor pool. Typically access was intergenerational. The skill requirements were industry specific, making steel could be learned nowhere but in a steel mill. There was some mobility only in maintenance, transportation, and clerical skills. The skill increments between jobs were so small that they could be learned on the job by substituting during illnesses and vacations. Hiring on at an unskilled level and advancing by seniority made sense both to management and to labor.

When the United Steelworkers began unionizing the industry in 1937, wage-rate inequalities between workers performing the same work in different plants became a serious issue in bargaining. Unionization of the entire industry came in the early 1940s and with it, as described above, the introduction of the Cooperative Wage Study which eliminated wage-rate differentials. The standard job classification system and accompanying wage-rate structure had a unique complexity. There were dozens of specialties eventually compressed into 32 job classes. The auto industry assembly plants by contrast had only assemblers plus some skilled maintenance personnel on each assembly line. In a steel mill, each division had many different occupations ranging through all job classes.

This set the stage for industry-wide wage bargaining. The management side was also interested in wage equality within regional markets in order to gain wage as well as price leadership. Price uniformity was more easily maintained when the biggest single input cost, wages, was also uniform. There was no product heterogeneity to justify price differentials and no technological heterogeneity to diverge production costs.

Industry-wide bargaining did not become a reality until 1955. Demand for steel was high and expected to remain so for many years. The dominance of U.S. Steel and oligopolistic pricing reinforced the union's strategic interests, though the latter was coming under pressure from imports.

In response to the perceived foreign threat, the US labor and management parties in the early 1970s negotiated a qualitatively new deal, the Experimental Negotiating Agreement (ENA). The union gave up the right to strike for an extended period in exchange for automatic wage increases—3% for productivity plus cost-of-living (COLA). The drop in steel demand and escalation of inflation in the 1970s' OPEC Crisis doomed the generous terms of the ENA to failure.

Eventually the ENA was not renewed and industry-wide bargaining itself was abandoned by the companies' refusal in 1985 to bargain any longer through the Coordinated Committee Steel Companies (CCSC).

Pattern bargaining was the new form of centralized bargaining begun in 1986, where a "lead company" was chosen by the union to set the industry pattern from one bargaining round to the next. This sustained a substantial degree of wage-and-labor-cost uniformity across the industry, mostly due to the union's efforts.

New Directions for Bargaining in the 1990s

Ironically, the implosion of U.S. steel companies in the 1980s saw leadership in the industry's collective bargaining approach coming from the union side in the "New Directions" bargaining policy led by Lynn Williams, the Canadian who became International President of the USWA, beginning in the 1993–1994 negotiations. This has ushered in an unprecedented level of labor–management cooperation in the steel industry and significant innovation in local workplace practices.

The Union summarized its principles as the Twelve Points of Light for the industry:

1. Full Union **Involvement** in all strategic decisions including the development of the company's business plan, acquisitions, design, and implementation of technological change, environmental issues, and access to all information pertaining to these issues. This included participation from the shop floor to union nominees on the Board of Directors.
2. **Employment security** guarantees.

3. **No concessions** in wages and benefits, and full and continuing access to the company's financial books and records.

4. **Neutrality** in union organizing campaigns and card check **recognition.**

5. Revitalized **apprenticeship** and **training** programs.

6. **Investment** commitments for plant modernization.

7. **Health care** cost reductions through managed care without penalties or cost shifting to employees.

8. Funding of **legacy costs** (unfunded retiree benefits).

9. **Corporate guarantees** to secure pensions and other benefits in case of company mergers, takeovers, sales or joint ventures, including successorship protection.

10. **Settlement of contracting out issues.**

11. **Joint public policy agenda** including rebuilding the nation's infrastructure, industrial investment, national health care, trade policy, and labor law reform.

12. Long-term (six-year) agreement.[10]

The above points constituted perhaps the most forward looking and ambitious bargaining program any union has ever tabled and successfully negotiated. However, the results were mixed due to external events and internal organizational dynamics.

Through the Cooperative Partnership Agreements (CPA) negotiated first in 1993, the United Steelworkers of America attempted to introduce its vision and model for more participative and productive work systems throughout the basic steel industry. The CPA structured joint labor–management planning, problem-solving, and decision-making processes at every level of the organization—shop floor, department, plant, division, and corporate, extending to union appointees for each company board of directors. It also prescribed participants from both the union and management at each level and included substantial training for all participants. In addition to improving relations and performance the USWA sought to increase its access to business information and influence over strategic as well as operational decision making.

Research was subsequently done to assess the effectiveness of the industry-wide contractual language and the centralized union-driven approach to workplace reform at the local level.[11]

The reasons for the US industry's problems included antiquated technology, an industry-wide bargaining structure that separated wage increases from productivity improvements, and a bureaucratic system of job classifications and work rules designed for an environment of market stability and the promotion of labor peace. This Taylorist system was composed of narrowly defined jobs, individual incentives, standardized procedures, strong managerial controls, and extreme specialization.[12]

Earlier, during the 1980 negotiations, the USWA and the industry attempted a new approach toward dealing with their productivity and quality problems. Much industry and public commentary at the time drew attention to Japanese Quality Circles and other team-based problem-solving practices. The new effort in steel was called Labor Management Participation Teams (LMPTs) and a two-page LMPT Experimental Agreement outlined a set of organizing principles.

While the Experimental Agreement was nationally negotiated, recognition was given to the fact that many of the industry's productivity and quality problems existed at the plant level in the organization, and process improvement through widespread workforce participation was required to solve them. Teams typically met weekly to work on problems ranging from quality and productivity to safety and the work environment, and union members along with their supervisors received up to 40 hours of training in subjects such as problem solving, statistical process control, conflict resolution, meeting skills, and group dynamics. Some companies also had plant- and department-level union–management committees to administer the team process and solve problems at a higher level, although the emphasis of the LMPT processes was at the team or shop floor level.

Later, the Cooperative Partnership Agreements attempted to provide a framework for: (1) joint decision making at all levels of the corporation, from the shop floor to the board of directors; (2) full and continuing access to business plans, records, and information including products, pricing, markets, capital spending, cash flow, finance, mergers, acquisitions, joint ventures, and new facilities; (3) jointly implementing new work systems and technology; (4) comprehensive education and training to provide the skills necessary for effective problem solving and participation. Specifically, improvements were targeted in quality, service, productivity, competitiveness, profitability, and safety. The agreement also advanced the goal to make the workplace "more equitable, less authoritarian, and less stressful."[13]

Finally, the agreement established a joint process for workplace redesign including the implementation of self-directed work teams and a shifting of responsibility for daily operations, planning, scheduling, and administration from supervision to bargaining unit members. It also outlined a process for the implementation of technological change including advance notice and information. If the parties were successful in implementing workplace redesign, then the union received joint decision-making authority over the effects of any technological change including the number and types of jobs required by the new technology, the skill and training requirements, the inclusion of new jobs in the bargaining unit, new work rules or operating procedures, and any health, safety, or environmental initiatives.

Further, this top-down contractual approach to industry-wide workplace reform provides an alternative to both the company- or plant-specific patterns of diffusion of high performance and participative work systems seen in most other industries, and to the earlier LMPT efforts in the steel industry itself.

The CPA called for extensive sharing of business information with the union. The majority of local unions reported receiving information on markets (80%), business plan development (75%), competition (70%), and financial performance (62%). However, information on long-term strategy (40%) was less forthcoming. Workplace redesign efforts had taken place in 42% of the plants, yet only 4% had attempted the provisions for a Joint Technology Change Program.[14] The experience of the USWA in implementing Cooperative Partnership Agreements across the US basic steel industry has been mixed. The agreements appear to have generally increased the sharing of business information and to a somewhat lesser degree many local unions reported positive results in the areas of quality, safety, decision-making, productivity, and cost reduction. However, in most cases the contractual provisions for CPA implementation have not been completely fulfilled. In particular, training was far less extensive than anticipated by the agreements, and many plants lack the departmental joint structures and team efforts on the shop floor. Further, provisions for joint workplace redesign and the union involvement in technological change have not been extensively implemented. Comparisons of successful and unsuccessful CPA implementation show significant differences in the extent of training, sharing of information, development of mid-level structures to support shop-floor teams, and managerial involvement. When these components were in place, the CPAs provided positive results for the locals unions involved. Further, three main barriers to more complete implementation of the Cooperative Partnership Agreements were identified. The first was managerial resistance, the second was the resources and priority given to the CPA by the International union, and the third was the nature of the agreement itself.

A challenge for the USWA with the CPA was both to get a commitment through collective bargaining that the parties could not walk away from, and to ensure the resources would be made available to implement the agreement effectively.

These mixed results on the CPA should not be interpreted as failure by the USWA. Rather they are a glass half full, diffusing workplace reform at a rate comparable or better than diffusion in other industries.[15] It should also be said that there was a level of reluctance on the part of the local union leadership to accept all of the responsibilities that came with the information flow and involvement with the business side of the steel companies. They intuitively resisted the prospect of accountably for decisions that might be unpopular.

Negotiated Restructuring in the 21st Century

However, much of this effort on both sides of the table encountered major difficulties in the 2001 steel crisis and subsequent bankruptcy of most of the so-called re-constituted mills. The International President of the union, Leo Gerard, took over the mantle of union leadership for the industry by his dramatic new efforts with the International Steel Group (ISG) agreement described below.

The steel crisis of 2001 brought a new wave of restructuring to the steel industry. The basic industry in North America and Western Europe never recovered from the destabilization that followed the Asian Financial Crisis of 1997–98 and the surge of cheap and dumped imports from Asia and Eastern Europe which followed. The most dramatic development has been the USWA agreeing to large-scale consolidation of the industry, as well as new collective bargaining agreements with the new owners, particularly ISG. These new agreements are qualitatively different from their predecessors in the 1980s. The newest deals are instead agreements with the minimill management teams being brought in to now run the remaining integrated mills.

Union President Leo Gerard decided to create a revolution in the U.S. steel industry. In an audacious move, in the midst of bankruptcy proceedings involving pension rights, he decided that the Union would reorganize the basic steel industry by creating its own steel company, find new investors, and hire a new management. The new entity became for a time, the largest basic steel producer in the United States, the International Steel Group (ISG) comprising the former LTV Steel, Bethlehem, and National until later taken over in turn by ArcelorMittal.

The terms of the new agreement became a reference point for the whole industry. Implementation and administration of collective agreements always present at least as many challenges for the labor and management parties as the negotiation of the contract in the first place. It remains to be seen how the new agreements will play out in practice and over time. However, there is no doubt the USWA-ISG was the most dramatic change in collective agreements in the history of the industry.

The Role of the Local Union in Steel

In their classic study of the local union, Sayles and Strauss identified the central focus and core activities of local unions as contract administration, grievance handling, and bargaining of local issues under the direction of national and international unions. However, the position and role of local unions are being transformed by efforts to involve employees more directly in problem solving, decision making, and improvements in workplace operations. Many of the principles guiding the structure and governance of local unions that grew up under the mass production Wagner New Deal model of industrial relations had to be changed. The changes are not easy and often do not come about without considerable political debate and internal and local–national union conflict.[16]

The most important change in the past 50 years has been an expansion of direct local union involvement in firm governance and management, challenging the boundaries of managerial authority established by the Wagnerist industrial relations system. This includes development of business strategy, new investments, product development, choice and introduction of new technology, training, job design, quality assurance, subcontracting, business planning, supplier selection, and work reorganization. Unions have also demonstrated their ability to add value on the shop floor by providing leadership, internal organization, and networking that brings effective coordination and implementation capability.

Other research has traced the differential levels of involvement of unions in the restructuring of the UK and German steel industries and the implications of such involvement on outcomes. The relative lack of institutions to allow for union involvement in restructuring decisions in the UK steel industry is contrasted with the more proactive responses by the German IG Metall union enabled by a much richer set of institutions in the German steel industry.[17]

For these reasons the forms of employment relations and contractual rules were critical to the evolution and performance of the industry. It is a deeper story than simply the aggregate capital investment and macrolevel labor market indicators relied on by Mangum and McNabb.

In Europe and Japan, stable, secure job tenure encourages reliance on internal labor market practices and creates a strong incentive for employers to invest in and support training activities. Co-determination rules arising from national labor laws confer a degree of authority on shop floor workers who, possessing the necessary skills, authority and long experience, intervene actively to solve operational problems as they arise. Their more extensive training makes it less likely that such problems will arise in the first place. The resulting "industrial culture" is common to both machine producers and users in Germany and facilitates the process of user–producer interaction leading to highly successful machinery design and use. Furthermore, the characteristics of the German workplace come to be reflected in German-designed technologies that assume a well-trained, highly skilled, relatively autonomous operator who will remain associated with the machine for a long period of time. The fact that such conditions do not exist to nearly the same degree in the North

American users' plants—and the resulting institutional distance between German machine tool producers and their North American users—is ultimately responsible for the implementation difficulties that have been documented.

Pension and Legacy Costs

Pensions are perhaps too appealing an item in labor management negotiations in heavy industry. There is a temptation to respond to a high union wage demand by tabling a counter proposal with a modest wage increase but a big move on the pension. It saves both sides of the table the trouble of struggling through the complex processes and sociology of high performance workplaces. It sells well with the union membership. It is appealing for the CEO as well because the costs are amortized over 15 years and he knows he is not going to be around when the bill comes in.

There are two problems with this. First, over time what happens is that the pension plan takes on the whole weight of the economic adjustment process. This is more than it can reasonably be expected to bear. Secondly, there are jurisdictions such as the USA and parts of Canada that have a Pension Benefits Guarantee Fund (PBGF) as an insurance device to guarantee workers benefits in cases of companies going bankrupt. The availability of this mechanism however can become a moral hazard for management. If a steel company blows out the fund by offloading its obligations, then someone is going to ask the moral hazard question and employers may wind up being told to self-insure.

It is not too much to say that, unlike in Europe or Japan where industrial policy imperatives have been the key drivers, in North America it has been legacy costs that have been the fundamental determinants of restructuring in the steel industry. The issues of pensions, particularly with regard to distressed companies and industries, raise fundamental questions about workers' rights, industrial restructuring, and regulation of corporations. Previously little known agencies such as the Pension Benefits Guarantee Fund (PBGF) in Ontario and the Pension Benefits Guarantee Corporation (PBGC) in the United States suddenly arrive on the front pages of the newspaper as central actors in determinations of whether companies are considered viable or not.

Pension and bankruptcy problems in the US industry 20 years ago foreshadowed problems experienced subsequently. Bankruptcy proceedings under Chapter 11 of the bankruptcy legislation in the US have in fact served as the vanguard for workers rights and restructuring disputes in much of industrial America. The leading study of the relationship between industrial restructuring and pensions in the US concludes that pension legislation and regulation simply does not provide adequate protection from economic imperatives because there is no general regulatory framework through which competing claims can be resolved and rationalized. The gap is the absence of an industrial policy, both for steel and more broadly for the economy as a whole.[18]

The US Employee Retirement Income Security Act (ERISA) of 1974 was designed to ensure that employers' unilateral or negotiated promises of retirement income were honored and legally protected. The act did not require employers to offer pension plans as part of their employment benefits or to offer a certain level of benefits. What the Act did was provide workers with legal rights to promised benefits while ensuring that the vesting, funding, and management of private pension plans meet a mandated minimum standard of operation. However, as the steel industry cases have made graphically clear, at root of what is in dispute is the appropriate balance between economic imperatives of corporate restructuring and corporations' obligations to their workers and society.

The 1986 bankruptcy of LTV was the watershed event in the struggle over the pension rights of steelworkers and corporate restructuring in the steel industry. LTV Corporation declared Chapter 11 bankruptcy in an attempt to shift its unfunded pension obligations to the federal government's PBGC.

At the time, LTV blamed its situation on the precipitous fall in domestic demand in the 1980s, the rise in steel imports and the lack of government policy to facilitate restructuring. Through Chapter 11, the company continued to operate and in the late 1980s regained profitability. The cost advantage of going into bankruptcy, particularly the transfer of unfunded pension obligations to the PBGC, tax concessions granted to the whole industry by the federal government and sustained improvements in productive efficiency and labor productivity through the bankruptcy proceeding significantly improved LTV's competitive position vis-à-vis other steel companies. The USWA, on behalf of current and former employees (pensioners), argued that retirees deserved more than the PBGC's mandated and guaranteed minimum monthly pension benefits. The PBGC claimed that the company could still pay most of its pension liability if it were to gain more concessions from union members. To that end, the PBGC applied to the courts to return most of LTV's pension plan assets to the company. It lost in federal bankruptcy court but won in the Supreme Court. The precedent of the LTV case has become a temptation for the whole steel industry to reduce costs by dumping off their pension obligations.

The company, notwithstanding an operating profit at the time had $4B in debt, some $2.3B of that related to pension obligations. Its pensions were chronically underfunded, unable to meet current or actuarially forecast, contractually agreed-to pension obligations. At the time, the LTV situation was viewed as the classic endgame for declining industries, the inevitable result of management accepting unrealistic wage demands in return for continuity of production. Unions were seen to have significant power in the short run because management was caught without investment options to go elsewhere. By this logic, union wage demands drove LTV into bankruptcy. The single and simplistic metric to support this theory is the often-cited, then as now, average wage for steelworkers versus the average wage for manufacturing or the economy as a whole. An equally if not more plausible explanation is that three combined circumstances in the 1980s were at play: the rationalization and restructuring of production capacity; the use of early retirement pensions by the company as the central means of reducing its workforce; and the rapid penetration of imports into the domestic market. LTV had consolidated Jones and Laughlin, Youngstown Sheet and Tube, and National Steel, reducing its combined workforce from 85,000 to 20,000. Rationalization added enormous long-term debt, largely the result of having to shed so much labor in so short a period of time, to an already indebted company that had absorbed the debts of the separate companies. It can be argued that LTV deliberately used its pension plans as the means of shedding excess labor through early retirement and plant closing pension benefits, believing that only this strategy would mollify the union's objections to the massive rationalization of employment. LTV acted legally in pursuing its merger strategy, but it did so knowing that the pension costs of this strategy might ultimately be borne by the PBGC.[19]

This is the case for considering steel company actions as strategic bankruptcy, for LTV then and other companies more recently. In effect, this strategy turns the public agency (PBGF or PBGC) into a creditor of the company. In the process, the LTV case and other current cases risks bringing the public agency to the brink of bankruptcy.

Bankruptcy proceedings are a very crude instrument for dealing with the chronic and complex issues facing the steel industry. Where does one find a decision rule to be fair to the secured creditors and sensitive to the circumstances of the company and its workers and retirees? An industrial policy or corporate restructuring policy might provide such a guide. In their absence, a judge is left to make policy in the context of competing interpretations of the logic of bankruptcy.

Today, the pension side of the legacy cost problem for steel companies has been brought under reasonably manageable control, though not without major controversy and radical redrawing of several of the major pension plans. The other side of the legacy cost issue—health benefits—now constitutes arguably a more daunting problem and will occupy major attention in steel industry bargaining for the foreseeable future.

Steel Unions and Labor Agency

Trade unions have been an ambient part of the economic and cultural environment of the steel industry around the world. They will continue to be so but what actual role they play in the globalized industry is very uncertain.

Academics refer to this as the question of labor agency. At the most general level, there are three potential things steel workers and unions can do. All of these involve having a "voice" function. Unions can advocate for employees around wage, benefit, and safety issues within the workplace, as reflected in their labor agreements. They can be an active voice in broader industry and environmental issues as membership organizations. Where unions are nonexistent or marginal, workers can merge with broader social groups as collective community voices in economic, social, and environmental affairs.

As revealed in the story told in this chapter, the role of unions in the steel industry is not the medieval siege between fixed positions that is often offered up in presentations by the media. Steel industry industrial relations have a lot of moving parts. We have seen, for instance, over 30 years of struggles in the US industry where government regulatory powers, specifically in the pension and benefits area, resulted in the United Steelworkers gaining strategic power outside the workplace as a party in the restructuring of U.S. steel companies.

Looking forward, the importance of environmental regulation is now the lead driver in the industry's technology investments for the coming decade. The intersection of union and civil society alliances will be a key determinant of the industry's horizon and prospects.

CHAPTER 7

Steel Trade Flows and Trade Disputes

Steel along with agriculture were the two industries most frequently afflicted by trade disputes in the postwar global trading system of the General Agreement on Tariffs and Trade (GATT). Business news in the 1970s and 1980s was dominated by trade conflicts between U.S. steel producers and Japan. Disputes with Europe were also common.

Why do steel trade conflicts get so much air time and drama?

Steel has been and remains a very cyclical industry. It makes an intermediate product. Therefore, decisions of others impacting things such as auto demand or decisions on capital investment in construction hugely affect the net demand for steel. These normal business cycle factors have in the past, however, been accentuated by movements in steel trade that can suddenly bankrupt billion dollar companies and throw thousands of workers on the street. Downturns in the economy have often been accompanied by dumping of imports, further depressing steel prices to destructive levels and wreaking havoc with capacity utilization, viable pricing, and layoffs. Dumped products and wild swings in trade flows can quickly turn the finances of capital-intensive industries like steel upside down.

There is some recent evidence to indicate that the globalization of steel has had a dampening impact on these destructive tendencies, that is, the Eastern European operations of a global steel company will not be allowed to put its North American operations into bankruptcy by dumping. However, there continue to be ongoing steel trade disputes, not least because of a slowing global economy.

The chapter begins by getting a base line for trade disputes—what is dumping? We then look at the changing economic geography of the global industry to gain a perspective on steel trade flows. The next section looks at steel trade issues and the Great Recession. The place of China, the elephant in the room, is then discussed. The chapter concludes with a discussion of steel and the state, that is, how governments are still playing a major role in how capital decisions and ultimately how steel trade flows unfold in the world economy.

Dumping: What It Is, Why People Do It

The epicenter of steel trade disputes, particularly dumping cases, has been the International Trade Commission (ITC) in Washington. It is important to understand that the ITC is a domestic, not an international agency. However, US legal definitions and policy decisions rebound around the global steel industry.

Under the provisions of the US Antidumping Act of 1921, the primary definition of dumping was export sales at a price below that of sales in the home market. Economists generally adhered to this criterion, defining dumping as price discrimination between national markets and explaining it with familiar theories of monopolistic behavior. However, the 1921 Act also included a provision to be invoked in the absence of comparable sales data for foreign markets. In such instances, dumping was said to occur when export prices failed to cover a statutory measure of foreign producers' production costs.[1]

Beyond the legal definition, there is the underlying question of why people do it if it means they are selling below costs. Sometimes it is because they are trying to force entry into a new market and establish a long-term position. In times of economic downturns, they are trying to keep minimum production running in order to cover base costs of operations. Particularly in emerging countries, there is pressure to earn foreign currency to be used for other purposes.

The 1974 US Trade Act and 1979 Trade Agreements Act further broadened the applicability of these constructed value provisions. As dumping allegations increasingly have come to revolve around the relation of prices to production costs, the issues have extended beyond reasons for price discrimination to encompass also the motivation for sales at prices that fail to cover costs.

The avowed purpose of the 1921 Act was to deter predatory pricing in international trade in order to prevent foreign monopolization of domestic markets. Its provisions, as incorporated into the 1930 Tariff, remained little changed until the 1950s. The Secretary of the Treasury was to investigate dumping complaints by comparing US import prices with the "fair value" of imports. On finding that fair value exceeded US import prices, Treasury was to calculate the difference, known as the dumping margin. Subsequently and with a different methodology, if there was evidence supporting a finding of material injury to US producers then Treasury was empowered to assess an antidumping duty. Measurement of US import prices was straightforward: the FOB factory sales price could be used except when the transaction between foreign supplier and US purchaser was not at arm's length, in which case the US market price, net of import charges and costs of transportation and preparation for the market, could be substituted.

From the law's inception, the calculation of fair value was ambiguous, since the concept was not defined by statute. From 1921 to 1954, Treasury used as a standard for fair value a commodity's foreign market value or, in its absence, constructed value. Foreign market value was a transaction price, preferably observed in the exporter's home market but otherwise in third markets.

Constructed value was a complex measure made up of allowances for production costs, costs of preparing the good for shipment, and statutory minima for general expenses and profits. Before 1955, Treasury calculations of fair value and foreign market value rarely proved problematic. Most dumping cases simply were disposed of on the grounds that injury was absent or on the acceptance of price assurances. In 1954, however, an amendment to the Antidumping Act assigned responsibility for determining injury to the Tariff Commission and instructed that injury decisions be deferred pending the Treasury ruling that dumping was present, thereby subjecting the Treasury decision to public scrutiny. Repeatedly, Treasury was forced to revise its procedures as new complications arose. On several occasions between 1958 and 1974, antidumping regulations were modified to bring them into conformance with established practice.

The amendments to the Antidumping Act contained in the Trade Act of 1974 concluded this process of revision. Of greatest consequence was Section 205(b), which defined new circumstances under which the constructed value criterion could be substituted for foreign market value. In instances where sales "over an extended period of time and in substantial quantities" were made in the foreign producer's home market at prices below costs of production, those foreign market prices were to be disregarded and constructed value calculations were to be substituted. In spite of ambiguity about the meaning of "an extended period" and "substantial quantities," this revision of the law represented a significant shift in the design of US antidumping policies from an emphasis on dumping as price discrimination to an emphasis on dumping as sales below cost.

In the 1980s and 1990s, the majority of steel trade cases by US producers were upheld regarding dumping and injury findings. Though, for all the reasons discussed in prior chapters, this did not prevent the painful restructuring of the industry. The related issues of government subsidies that distort trade flows were significantly eased by the introduction of a clearer and more accepted definition of what constituted a subsidy. But there remains a steady stream of dumping allegations.

On the injury side, we may be in the midst of a policy shift. Negative determinations by the ITC on injury determinations are relatively rare. However, three recent ITC cases rejected claims that steel product imports were hurting the US industry: steel wheels from China, refrigerators from Mexico, and galvanized wire from Mexico and China. A majority of cases that get to final determination are affirmative. The recent cases may amount to a game changer. It may cause domestic producers to think twice about making such filings. The expense of filing cases and uncertainty over outcomes raise the stakes for potential litigants. If you file and lose, it is a big loss in legal costs, management time and a significant loss for the market position of the industry.

The Economic Geography of Global Steel

Understanding the shifting economic geography of steel is critical for gaining a perspective on the emerging global industry, including its relevance to steel trade issues. Where are the major producers located, what are the balances and imbalances of demand and supply, what is the balance of trade, and what are the dynamics of growth?

To repeat, it is important to understand that while there are global companies there is not one global market in steel. There are three key regional markets: North America, Europe, and Asia.

Steel Use Around the World

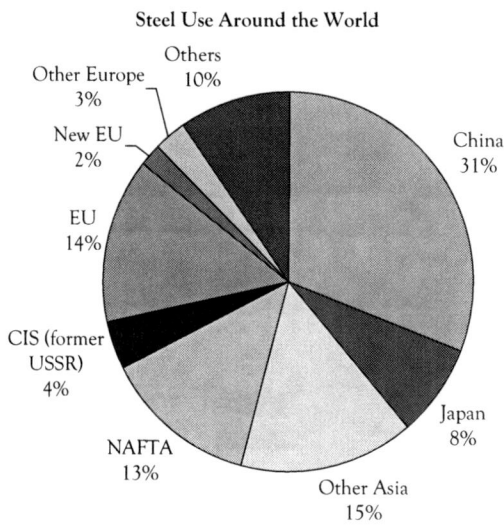

Steel Production Around the World

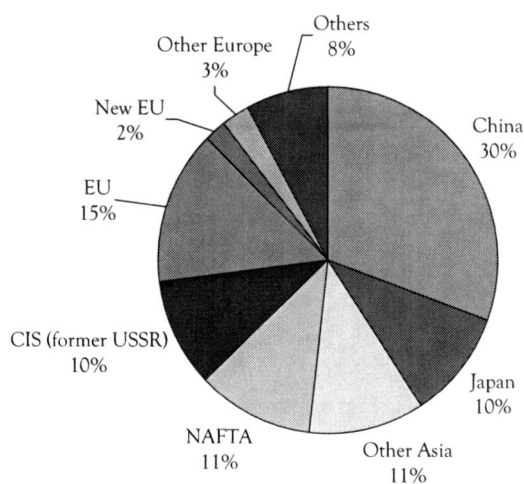

At first glance the global steel industry appears to be in approximate balance between production and consumption.

The big markets of NAFTA, the European Union, and even China seem to be reasonably well balanced in terms of demand (consumption) and supply (production).

However, as in many things, the devil is in the detail. The macro economic view does not readily disclose the subtleties of market dynamics and the underlying tensions in steel trade and global production flows.

Key Regional Markets

As mentioned above, there are three key regional markets in global steel: North America, Europe, and Asia. The numbers should be considered against a total world capacity of 1,351 millions of tons as of 2010.

Ten Largest Steel-Producing Countries (Millions Tons)

Rank	Country	Steel production
1	China	626.7
2	Japan	109.6
3	USA	80.6
4	Russia	67.0
5	India	66.8
6	South Korea	58.5
7	Germany	43.8
8	Ukraine	33.6
9	Brazil	32.8
10	Turkey	29.0

China is a story unto itself. The remaining key markets of Asia, Europe, and North America are roughly comparable in size.

Steel Trade Balances

There is a fundamental asymmetry in global steel markets. Similar to the auto industry, there is a North American market that has significantly more imports than exports and an Asian market with significantly greater exports than imports, led by Japan and China.

Steel Trade Balances (2010 Million Tons)

Region	Exports	Imports	Net trade balance (imports–exports)
North America	24,163	40,238	–16,075
European Union	134,633	125,115	+9,518
Asia	132,325	114,235	+18,090
China	41,646	17,181	+24,465
Japan	42,735	4,440	+38,295

The trade imbalance in North America has other serious implications. If we even take the high-end productivity goal of steel companies at 1200–1500 tons shipped per employee per year, then this translates into about 15,000 jobs missing. With the standard multiplier effect for steel of 6x, then nearly 100,000 jobs are missing in North America compared with what the employment situation would be if there was balanced trade in steel. That is an employment outcome that would change people's perception of the importance and future of the industry among other things.

Steel and the Great Recession

As a result of the global economic crisis, NAFTA steel production declines in 2009 were larger than those in other regions of the world.

Global crude steel production
2009 YTD vs. 2008 % change

N. America: –45.1   Turkey: –13.5
Canada: –51.9   EU27: –39.3   Russia: –26.8
U.S.:   –47.0   Ukraine: –31.9
Mexico: –29.5

                            Asia: –2.2
                            Japan: –34.0
                            S.Korea: –14.9
                            China: +7.5
S. America: –30.3   India: +1.6
Brazil: –31.4

Global production: –16.4
Excluding China: –30.9

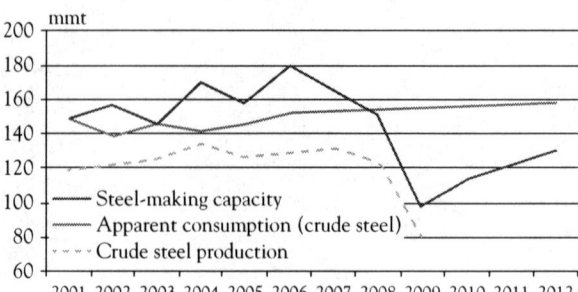

NAFTA Capacity and Demand Until 2012

As a result, the NAFTA Steel Trade Balance with the rest of the world (ROW) going forward is in a fundamentally different situation. The steel balance in the region has shifted from one where it was historically steel short (requiring imports), to one where it now has capacity to increase production for domestic and export consumption, without the need for imports. Given the anticipated postcrisis steel consumption growth in NAFTA and ROW, this is a significant opportunity for NAFTA steel producers.

Steel Trade Disputes

As background to understanding trade disputes, certain structural features of the steel industry can have a significant influence on trade flows. Steel making is capital intensive and involves relatively high fixed costs. Consequently, there is an incentive for producers with significant excess capacity to increase production to spread fixed costs over a greater volume of production. However, there is a countervailing incentive to align production with demand in a steel maker's domestic market. Excess supply in the "home market" can result in pricing instability, which can negatively impact returns.

International trade agreements are also important. The emergence of the NAFTA steel market has resulted in a situation where the home market for steel makers in the region is the NAFTA market. However, global excess capacity and the emergence of China and other countries as major steel exporters introduced significant challenges for steel makers here.

The combined effect of global overcapacity and the incentive to maintain high production levels creates an incentive for steel makers located in markets with significant excess capacity to increase production for export markets. This allows steel makers to act in a manner that promotes pricing stability in the home market while increasing capacity utilization by selling into export markets.

This situation may be further exacerbated in situations where governments adopt policies that influence production decisions and/or confer production or export subsidies on steel products.

The combination of overcapacity and government involvement has resulted in widespread dumping and subsidization of steel products on export markets.

Over the past decade China has emerged as a major driver of international trade disputes. For example, since joining the World Trade Organization (WTO) in 2001, Chinese government measures affecting a wide variety of industries have been challenged before the WTO. Other WTO members have brought challenges against measures affecting auto parts, financial services, intellectual property, taxation, and technology products.

The lengthy list of Chinese steel products found to have been dumped and/or subsidized in export markets illustrates that the structural features of the steel industry combined with government influence and support have resulted in a pattern of dumping and subsidization in export markets.

Most recently, the United States, Mexico, and the European Community have initiated WTO dispute settlement proceedings regarding the Chinese government's system of export restraints affecting raw materials. Canada is an active third-party participant to the proceedings. It is alleged that China maintains a system that restrains exports of raw material inputs used in the production of finished goods such as steel. The export restraints are alleged to raise world prices while lowering Chinese domestic prices for key steelmaking inputs such as coke.[2]

The Effects of Dumping in the Domestic Market

Unfair trade practices are often the result of asymmetric market access and economic distortions in the exporter's home market. Dumping

Duties Imposed on Chinese Steel Products

Product	Canada Dumped	Canada Subsidized	United States[3] Dumped	United States[3] Subsidized	Mexico[4] Dumped	Mexico[4] Subsidized
Hot-rolled sheet	X		X			
Plate	X		X		X	
OCTG	X	X				
Seamless OCTG	X	X				
Standard pipe	X	X	X	X		
Carbon steel butt-weld pipe fittings			X			
Drill pipe			X	X		
Steel concrete reinforcing bar			X			
Steel nails			X	X	X	
Light-walled rectangular pipe and tube			X	X		
Steel wire garment hangers			X			
Circular welded carbon quality steel line pipe			X	X		
Circular welded austenitic stainless pressure pipe			X	X		
Steel threaded rod			X			
Welded steel chains					X	
Welded carbon steel pipe					X	
Seamless Steel Pipe					X	
Steel bolts					X	
Steel valves					X	

and subsidization cause negative economic and social consequences on affected communities.

The existence of nonmarket influences and government support of the Chinese steel industry serve to increase Chinese exports and cause distortions in the importing country market. The economic consequences of dumping or subsidization from the importing country's perspective are illustrated in the following graph.

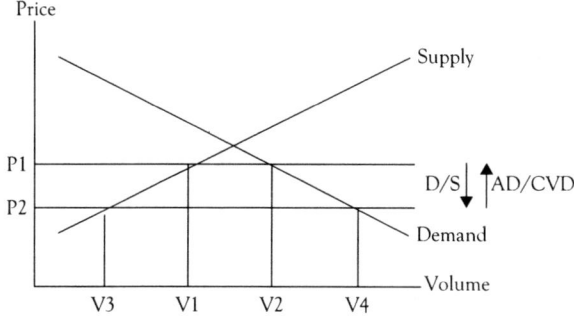

Effects of Dumping/Subsidization
in the Domestic Market

P1: World price
P2: Dumped or subsidized price

V1: Domestic consumption from domestic
     production with fair trade
V2: Total domestic consumption with fair trade
V3: Domestic consumption from domestic
     production with dumping and/or subsidization
V4: Total domestic consumption with dumping
     and/or subsidization

Government support and other structural factors allow exporters to lower their selling price from the world price (P1) to the dumped and/or subsidized price (P2). The impact on domestic producers in the importing country is decreased volumes sold in the home market (domestic share drops from V1 to V3). Exports at unfairly traded prices experience a corresponding increase (V2 to V4). D/S stands for dumped or subsidized prices. ADV is ad valorem duties applied.

Trade laws minimize the disruptive economic and social effects that unfairly priced imports have on established communities by restoring market equilibrium.[5] The imposition of antidumping and/or countervailing duties on dumped or subsidized exports restore production to undistorted levels by offsetting the effects of the dumping and/or subsidization.

Trade Liberalization

The negotiation of trade agreements involves the balancing of concessions and opportunities by the parties to the agreement. In principle, each agreement should be assessed to determine whether, as a result of a given agreement, the industry is better off than it would be without the agreement. The existence of effective trade remedy laws contributes to trade liberalization by providing a mechanism to address specific concerns about the potential negative effects of unfair trade, while allowing for broader trade liberalization.

From a trade policy perspective, trade liberalizing agreements should increase the overall size of the market available to domestic producers. This involves an assessment of the access granted to the domestic market in exchange for improved or expanded access for domestic producers in export markets. Providing trading partners with expanded opportunities in the local market is one-half of the equation. The net benefit of a given agreement can only be understood by examining whether the increased access offered to the domestic market also affords local producers with an equivalent or greater opportunity in export markets.

As noted above, the NAFTA market has shifted to a steel "long" situation, which means that producers in the NAFTA countries have the ability to serve the needs of the NAFTA market as well as export markets.

The view of the industry is that to maintain a steel balance in the NAFTA region, there must be a public policy commitment to restore North American manufacturing as a foundation for economic growth and sustainable employment. They view China as in effect pursuing a mercantilist policy in violation of the content and spirit of the international trade rule regime. Furthermore, they warn about the risk that inequitable application of climate change policies will allow those with little to no regulatory burdens to in effect, engage in environmental steel dumping. The latter would both be trade-distorting and also prejudice the environmental and sustainable development objectives that the NAFTA steel producers themselves endorse.

The Impact of China

The huge expansion of Chinese steel capacity has raised concerns about the potential of dumping and a flood of Chinese imports. The current facts are that Chinese steel exports are a rather small amount of total

Chinese Exports of Steel Mill Products by Partner
March 2012

South Korea 19.4%
Thailand 5.3%
India 4.9%
Vietnam 4.7%
Philippines 3.9%
United States 3.0%
(8th in volume)
Other 58.9%

Total quantity:
4.8 million metric tons
Source: Global trade atlas

Source: Department of Commerce. International Trade Administration. May 2012.

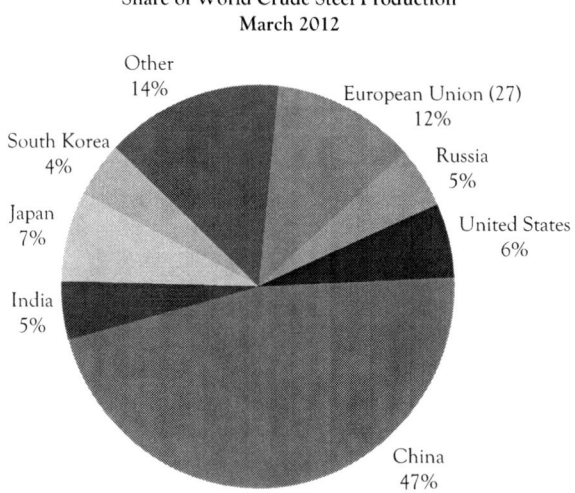

Share of World Crude Steel Production
March 2012

Other 14%
European Union (27) 12%
South Korea 4%
Russia 5%
Japan 7%
United States 6%
India 5%
China 47%

Total world production: 132.2 million metric tons
source: World steel association (formerly IISI)

Source: Department of Commerce. International Trade Administration. May 2012.

production, about 4%. However, the sheer size of the Chinese industry, approaching 800 M tons, can still mean serious tonnages on the order of 40–50 MTs or more are available for export. Therefore, it matters a lot where these Chinese exports are going. The graphs present some of the most recent relevant data points.

It may come as a surprise to many that most Chinese steel exports are going into the Asian market. For instance, China sends eight times the amount of steel to South Korea that it does to the USA. However, there are complex dynamics concerning China's presence in the Asian regional market. Japanese steel producers are increasingly concerned about shipments coming in from China, or even more that shipments to South Korea will result in the latter producers shipping to Japan in a second bounce effect.

The challenge for producers and policy makers is China's potential to directly upset this equilibrium through non-market-based behaviors. In addition, there is the problem of indirect trade in steel because of China's presence in manufacturing and further displacement of NAFTA manufacturing capacity.

China now represents almost half of global steel production. In the past 10 years, it has increased its crude steel production by over

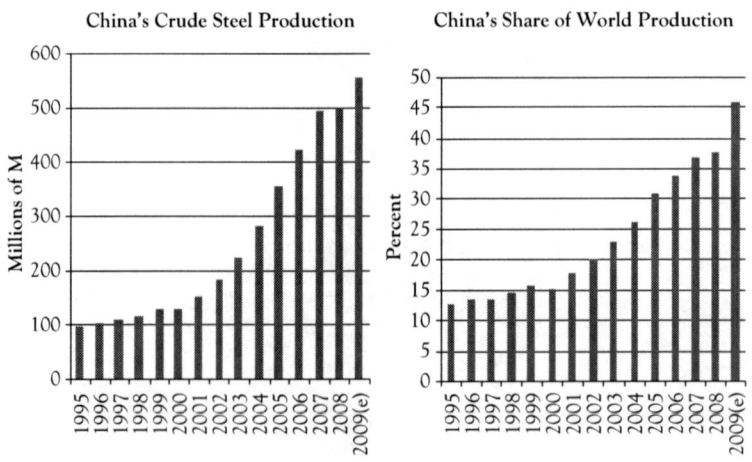

Source: World Steel Dynamics, Inside Track #77 (May 30, 2007), World Steel Dynamics, Truth & Consequences #44 (Nov. 15, 2007); World Steel Dynamics, Inside Track #87 (June 12, 2008); World Steel Dynamics, Truth & Consequences #50 (Feb. 9, 2009); World Steel Dynamics, Inside Trade #97 (Oct 2, 2009).

400 MTs, and increased its share of world production from 15% to almost 50%.

The best source of data and forecasts for global steel come from the OECD Steel Committee in Paris. Their most recent Steel Outlook is from December 2009.

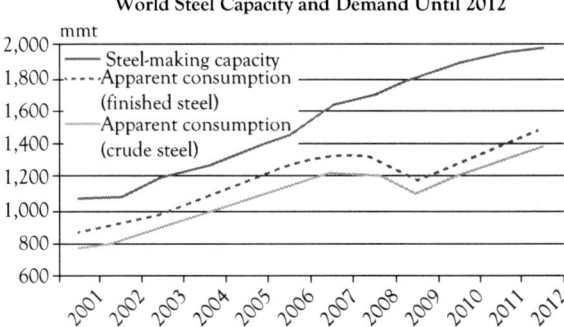

World Steel Capacity and Demand Until 2012

The OECD sees world trade as recovering over the next three years, led by the major non-OECD countries of the BRIC. Their projections suggest that world steelmaking capacity would rise from 1,806 million tons in 2009 to 1,986 million tons in 2012. World steel demand is expected to rebound in 2010, and grow by 6–7% per annum in 2011–2012 to reach a level near 1,500 million tons by the end of the period.

While there is uncertainty surrounding the outlook, it appears that the gap between world capacity and demand, which averaged approximately 216 million tons during 2000–2007, will widen to over 500 MTs. Such an overhang presents significant structural challenges to the industry. It raises questions about how the industry adjusts and what government policies might help manage the situation.

In the critical case of China, the OECD observes that capacity growth outstripped demand growth from 2002 to 2007 and it turned into a large exporter of steel in the latter part of that time period. By 2012, based on the OECD forecast, China will still have an excess of capacity in the range of 150 MTs, or approximately two times the size of the entire American steel industry.

Global steel companies are unanimous in the view that China's steel industry is firmly embedded in a powerful state–business nexus. Chinese

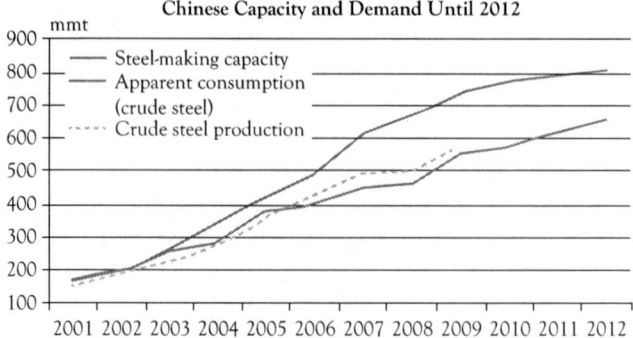

steel companies are not operating in competition based on the domestic market environment, but rather hold very close relations to government agencies on local, provincial, as well as central levels. Except for two enterprises, their top 20 steel corporations are state-owned on a majority basis. In sum, China is a nonmarket economy in steel. Its steel industry, a multilayered system of politico-business alliances can be summarized in the following schematic:

Chinese governments support their steel enterprises through a National Steel Policy and provincial/local actions. The broad array of mechanisms includes currency/capital market interventions, direct and

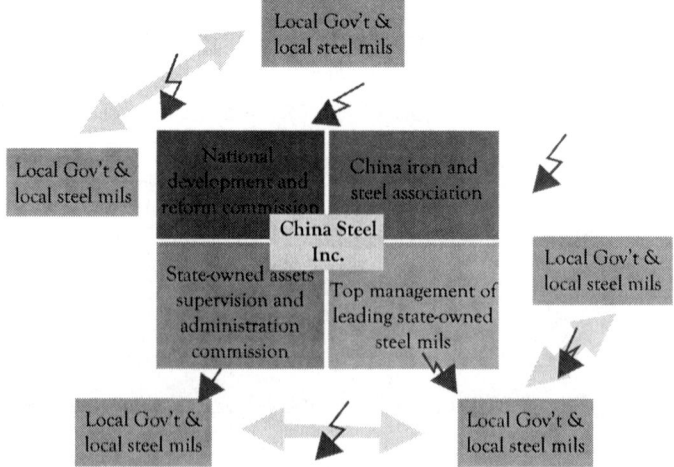

Source: Report prepared by THINK!DESK China Research & Consulting for EUROFER—the European Confederation of Iron and Steel Industries. January 2009.

indirect ownership, subsidies, quotas, import/export taxes/rebates, export targets, etc. Government intervention is provided across the entire supply chain, including energy, raw materials, steel, manufactured products, and more. The interventions are structured to artificially and selectively increase the competitiveness of Chinese goods, while concurrently increasing costs and decreasing relative competitiveness of other global players. China is no longer just a supplier of lower added-value steel products; instead, it has shifted to export more higher value-added materials encouraged by government.

For an outsider's viewpoint, the steel situation is strange as China lacks many natural advantages for steelmaking. China must import significant amounts of quality raw materials, at world prices, which represents the majority of their total production costs. Growth of coal-fired plants, limited supply of steel scrap, and less efficient and environmentally challenged mills reduces their competitive balance. The availability of cheap labor does not offset their real cost disadvantage, as steelmaking generally now requires less than 2 hours of labor per ton.

As a result, Chinese steel exports to NAFTA and to the European Union actually incur higher costs than those that arise for local producers supplying the local market. Netting out subsidies and the impact of government market interventions would prove the real 'market-based' cost structure of Chinese steel production to be substantially higher than officially reported numbers.

Comparative Steel Production Costs: China versus EU

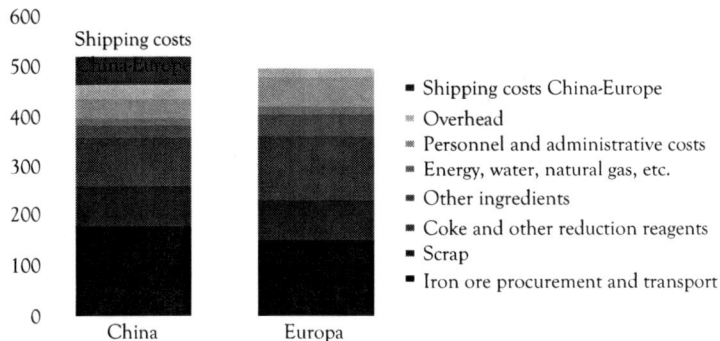

Source: Report prepared by THINK!DESK China Research & Consulting for EUROFER—the European Confederation of Iron and Steel Industries. January 2009.

Nobel Prize-winning economist Paul Krugman has recently said that global economic growth would be about 1.5 percentage points higher if China stopped restraining the value of its currency and running trade surpluses. He says that China's currency policy has a "depressing effect" on economic growth in the US, Europe, and Japan, as measured by gross domestic product. If China's currency, the yuan, were not undervalued, it would have a "significant" impact on the global recovery.

China also has a significant impact on manufacturing trade deficits and indirect steel trade deficits.

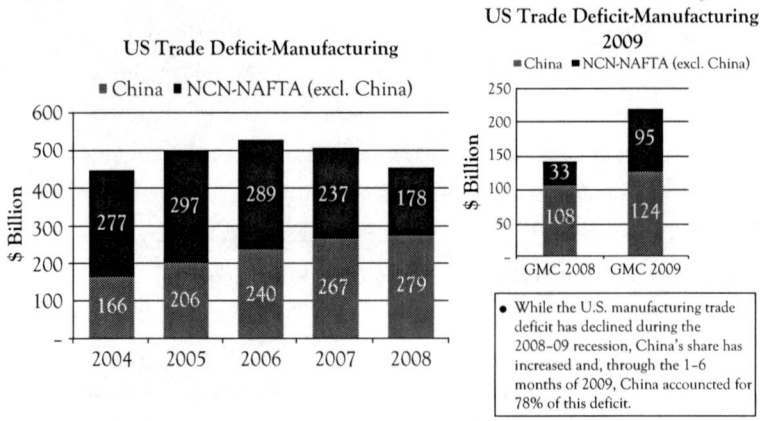

For steel in particular, the indirect steel deficit trade remains a major source of concern.

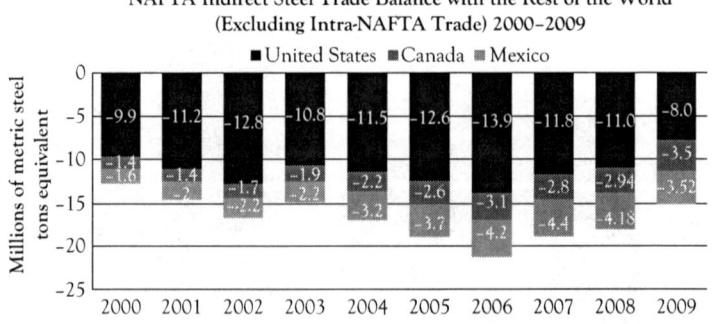

The question of China's role is a preoccupation for steel companies around the world. The crunch point may come in the next two years. China's steel consumption is expected to peak at 750 MTs in 2014.

At that point, there will be serious overcapacity in their industry and a restructuring scenario emerges. Will this excess capacity be put out into the world markets in the form of an export surge? The Chinese government opposes this and has taken some actions to restrict it happening. If this remains the case, we will witness another surprise in steel development in China—a painful process of industry restructuring.

Global Steel, the State, and Trade Flows

The steel industry, as discussed previously, has been globally consolidated to an unprecedented degree. One of the expected benefits of the consolidation was the dampening down of some of the key drivers of disruptive steel exports and dumping. Private corporate interests take precedence over nationalistic and mercantile agendas. As this book is being written in mid-2012 the evidence for a more benign era of international steel trade flows is at best mixed. During the recession, imports lagged but there are renewed fears that as things gradually improve, while China slows, there is a risk of a renewed import surge.

The general challenge of trade and its relationship to capital flows can be crystalized in the dilemma facing the largest steel company in the world. The biggest player in the industry, Arcelor Mittal, is at the center of this strategic story. It has been the lead consolidator internationally and it could reasonably be expected that, for instance, Arcelor Ukraine would not be allowed to put Arcelor USA out of business. Facilities would reduce production levels rather than dump product abroad.

In the immediate aftermath of the Financial Crisis, 2008–2009, there was some evidence for this policy thesis. The steep reductions in steel production and capacity utilization levels were not matched with as steep declines in market prices.

However, there was an upside to the positive expectations of consolidation. As the economy turned around, it was expected that companies could increase prices to catch the upward movement of the business cycle. However, the evidence of 2011–2012 does not match expectations. Steel companies tried to increase prices substantially, led by Arcelor, but these have not proved to be sustainable. Renewed import concerns and trade cases have re-emerged during the current slow recovery.

Trade Policy and Competition Policy
Inevitably Intersect

One of the puzzles about the steel industry is the issue of steel being a big industry at the national level, but a borderline player at the global level. Part of the answer to the puzzle is that steel, largely because of the weight of the product and the scale of efficient production, is produced largely, be it by integrated or by EAF producer, for regional markets. A large part of this in turn is because of manufacturing being oriented to geographic clusters.[6] It is these regional concentrations of the customer base that largely dictates the configuration of steel producers.

Meanwhile, the concentration in ownership in iron ore and coal producers is an outlier in the global economy and probably will undergo antitrust and other commercial law challenges in the coming years. European antitrust authorities are already seized of the issue.

The conclusion is that there will, for the foreseeable future, be a global steel market that is very much regionalized. It will, however, be brought into better balance through a combination of corporate, legal, and public policy motivations. In steel, as in auto, the North American market continues to be unique in its imbalance of imports and exports compared with both Europe and Asia.

Nation states are still important in steel. Among other things it impacts the evolution of steel industries, particularly in Asia. It is even a major challenge for the largest steel company in the world, Arcelor Mittal.

Steel and the State was a common point of discussion in policy debates and the academic literature 25 years ago. The issues of protectionism and dumping, public subsidies, and excess capacity issues were common questions in the media and trade regulatory decisions. In fact, in the later 1980s, a majority of steel capacity around the world had direct and indirect government ownership and subsidization writ large. Much of the liberalization of steel economics in the 1990s and carrying into this century was about the wind down of public ownership in steel and a major reduction of trade-distorting subsidies. The consolidation wave in steel in the past decade has been very much a case of the takeover of formerly public steel assets by private steel companies. This was key to the emergence of Arcelor Mittal but as well of the new presence of U.S. Steel in Eastern Europe.

Raising the issue of steel and the state here is not to go over that old ground one more time. It is to raise a current significant issue of steel and the state in the new environment of globalized steel. It is possible to say broadly that a major strategy of the new global steel companies is to be present in all key markets, particularly those of North America, Europe, and Asia. However, to implement that strategy requires access to those markets. The elephant in the room is the government of China, which will not allow majority foreign ownership of steel assets. For a comparison, one can see the importance of the role(s) of the state by considering how the global auto industry's expansion into Eastern Europe is so fundamentally different from the emerging auto industry in China. This anomaly is evident even to editorialists in the *Financial Times of London*. The challenge for global steel companies in implementing a truly global strategy has been well identified recently by their description of the strategic dilemma facing Arcelor Mittal.

Sometimes a company prays for a strong Chinese economy not so that it can sell more products to China, but so that the country will sell fewer products to the world. Take Arcelor Mittal. Its revenues rose by 20% last year to $94bn, and its crude steel output edged up to 91.9m tons. The distance separating Arcelor Mittal, the world's largest steelmaker, from second-placed Hebei Iron & Steel of China is impressive.

There is, however, a twist. Arcelor Mittal's founder Lakshmi Mittal puts a brave face on matters, but the truth is that his company produces and sells little steel in China. Beijing refuses to let foreign steelmakers take majority stakes in large Chinese companies, so Arcelor Mittal, which only makes about 7% of its steel in China, must rest content with two modest joint ventures.

For Arcelor Mittal, which produces about 6% of the world's steel, such trends are troubling. It is desperate to protect its global market share without suffering the price pressures that might emerge as a result of fast increasing Chinese production. Yet its Chinese rivals need to sell their steel somewhere.

The best hope is that China records buoyant economic growth this year. In principle, this would absorb much of the extra domestic steel output and, as a happy side effect and curb the appetite of Chinese companies to seize world market share from Arcelor Mittal. Not for nothing is Arcelor Mittal counting on 8% growth in China in 2012.

In Europe, the steel story is different but familiar to anyone who knows the auto sector story: structural overcapacity in the industry, and a backdrop of low long-term economic growth.

The *Financial Times*[7] has stated the strategic dilemma for Arcelor Mittal. Mr. Mittal became famous as the businessman who defied the received wisdom of the 1990s that steel was an industry on which the sun was setting after 150 years. By creating a group that now employs 260,000 people in more than 60 countries, Mittal showed how to make healthy profits out of steel.

But Arcelor Mittal is aware that European steel demand is unlikely to return for the foreseeable future to the levels seen before the 2008 financial crisis. With European plants accounting for about 40% of his company's output, this leaves Mr. Mittal with little choice but to reduce operations.

The process is already in motion. The closure in October of two blast furnaces in Belgium was the first such action since the company's 2006 creation out of the €26.9bn merger of Mittal Steel and Arcelor. Elsewhere, mills are idle in Spain and a voluntary redundancy scheme is being implemented at a Czech plant.

However, none of these actions gets to the heart of Arcelor Mittal's strategic problem. The 2006 merger was founded on the twin assumptions of an Asian hunger for steel, which the company could feed for decades to come, and a resilient European market primed for its high-tech products. The strategy was and is not completely misplaced. But it has suffered a blow from Europe's debt crisis, which has dramatically reduced demand for cars, construction, and other steel-buying sectors. At $10.1B, the company's earnings before interest, tax, depreciation, and amortization last year were less than half those of 2008. The steep decline over the past three years in Arcelor Mittal's share price was all but unavoidable.

Blocked in Asia and on the defensive in Europe, Arcelor Mittal, the steelmaker par excellence, now styles itself a "steel and mining company." Iron ore and coal account for less than 10% of its business, but mining is growing fast and makes a tidy profit. The company now has a specialist mining management team and reports its results as a separate segment.

Time will tell if this is a sensible response to Arcelor Mittal's challenges on the steel front. But digging raw materials out of the ground was not exactly what Mr. Mittal had in mind when he executed his famous 2006 merger.

Arcelor Mittal is by a large margin the biggest company in the global industry. This huge private corporation still faces its limits in the face of an assertive nation state. Further, it underlines that flows of investment capital are absolutely critical to steel trade flows for the future fortunes of the industry. And, that these flows function well short of the expectations of free markets.

CHAPTER 8

Outside Market Forces

Steel is an intermediate product. It is sold to people who make other things with it. Therefore, the nature and dynamics of steel-consuming industries are critical to understanding key drivers of steel companies and the industry as a whole. These outside market forces are unusually important for the directions and dynamics within steel, perhaps more so than if steel was a standalone industry.

For almost a 100 years, "the" key customer for steel has been the automotive industry. The auto–steel connection is so deeply embedded that it is fair to say that we would not have the auto or steel industry in the form that we have them, one without the other. For this reason, the chapter begins with a summary of some of the critical important developments and stages of automotive steel. The two industries co-evolved.

Developments of supply chains in automotive have led to new practices across manufacturing. In a simpler time, the price negotiations between U.S. Steel and GM used to determine the price of flat rolled for all of North American manufacturing. However, in recent years, the stretch out of auto supply chains has played a major part in transforming the steel industry. We look at the role of various agents along the supply chain.[1]

We also discuss the functioning of steel companies in local markets and the emergence of local steel economic clusters.

Does steel have a growth story? The future of the steel industry largely depends on the outlook for manufacturing in general. For some, steel is seen to be following the apparently inevitable decline of automotive and manufacturing in general.

Finally, because of the intimate linkages between steel producers and consuming industries, the industry is vulnerable to amplification of the normal business cycle swings that all businesses must manage. The so-called Bull Whip Effect (BWE) is considered in the context of the recent Great Recession. The latter is a special case of how outside market force impacts the industry.

The Auto Steel Story

Taken together, the steel and automotive industries constitute the largest manufacturing complex in the world. It is not possible to talk of one without the other. The basis of production economics in auto to this day turns on the all-steel auto body or "unibody" developed in the 1920s. Its linkages to auto manufacturing and assembly technology—a 75-year learning curve of skills, knowledge, and investment—is the fundamental reason why neither aluminum or plastic will replace steel in cars any time soon.

Henry Ford and E.G. Budd

The conventional wisdom of 20th-century business history states that the revolution at the Ford assembly line at Highland Park in 1913 gave rise to the mass production–consumption model of the modern industrial economy. Fordism is what academics call it.

The key was the introduction of the assembly line that reduced production time for a Model T from 12.5 h to 1.5 h. The resulting economies of scale allowed Henry Ford to reduce the price of the model T from $950 to $350 and facilitated his revolutionary offer to pay $5 per day to his auto workers so they could afford to buy his cars. The new mass consumer market was born.

In reality, Ford was more an innovator in outsourcing and logistics, not assembly production. The introduction of conveyors at Highland Park had their primary impact on the flow of components. They were not the primary pacers of production on the final assembly line. The pioneer of the mass production assembly line was not Henry Ford but E.G. Budd.

The paint stage had become the bottleneck of the new auto assembly lines. Converting from wood to steel bodies was Budd's solution. It has remained the key determinant of the economics of auto production to this day, through the modern unibody design used by all producers of high-volume automobiles.

Until the early 1920s, most cars had wood or wood/steel composite bodies. The automobile body was basically a wooden frame with steel sheets tacked onto it. The steel was used on the body but not for the frame. Body construction was slow and costly.

In the first decade of the century, it was estimated that it took 106 days to produce a sedan body from the lumber pile to finished product.

It now takes about 29 h. About 25% of that time was spent in the paint shop. Twenty-four operations were required to apply paint and varnish, involving 14 drying periods, each taking from six hours to a full day. There were literally acres of storage space at auto plants covered with automobile bodies in various stages of completion. Enamel required a baking temperature of 400°F, which was impossible for wood bodies. Lacquers were eventually developed that reduced drying time from 30 days to 3 days, but this was still not enough to keep up with demand.

Before being hired by Ford, Budd produced his first steel bodies for the Oakland motor car and then the first large introduction was for the Dodge brothers in 1915. Steel construction solved the fundamental bottleneck because unlike its wood predecessors the steel body could be completely dipped in enamel then baked dry, quickly.

Budd had an early interest in press-formed sheet steel. The new Dodge cars had a significant advantage in weight, strength, safety, and durability. However, they required a radical departure in automotive manufacturing practices: it required the highest-quality steel sheet at the absolutely maximum width. Most forming operations were eliminated. The new approach required an unprecedented use of stamping machines. A stamping press is a metalworking machine tool used to shape or cut metal by changing its shape with a die. It also forced the development of new welding techniques.

The Budd body was put together by welding together four main subunits: the outer body shell, the inside frame, the seat supports, and the doors. These in turn, were each built up by the welding of many smaller parts.

By 1927, the Budd Company developed what we now call the unibody, a structure in which the strength factor was reversed since the outside steel rather than the inside frame provided rigidity and support. Actually there was little inside framing, since the side sills had become obsolete and the assembled body was attached directly to the chassis.

With a series of overlapping patents, Budd defined the pressing and welding techniques required to construct a car body from a three-dimensional "jigsaw" of sheet steel panels. The all-steel body gave some immediate and important product benefits: strength, stiffness, greater design freedom, suitability for painting, greater precision, enclosed car bodies; and was much better suited to continuous manufacturing techniques with a significant fall in unit costs under high-volume production.

Steel Changes Automotive

As stated, most early cars were made with wood frames and steel sheet exterior body panels. As a material, wood never fitted in with mass production auto manufacturing. Critical items such as the main pillar-post were intricate composites requiring dozens of specialized and time-consuming operations. It had to be cut from solid hardwood, and since it suspended the door and held locks and other hardware, it was honeycombed with indentations, holes, and notches, many of which were sheathed with metal. As a result, before 1935 and the introduction of the complete steel autobody, the assembly line was not widely used in automobile body construction. Assembly line methods could only be used if the wooden frame was eliminated and the body made entirely of steel. This would allow manufacturers to stamp out the various parts such as fenders, hood, and so on, in huge presses and then weld them together to form the body.

Following the Budd innovations, the all-steel body became the single most important element in vehicle design. The manufacturing processes required for all-steel bodies became the core investment for vehicle manufacturers. Even now, in contemporary high-volume car plants, it is the press shop, welding lines, and paint shop, all required as a direct result of using all-steel bodies, which account for the majority of investments required. Equally, for each model produced it is the tools and dies to make the body that account for the largest investments.

The 1920s had also seen the enhancement of automotive design and styling, facilitated by steel. The original design—a box sitting on wheels—had been little changed. Developments in aircraft design in World War I and the science of fluid mechanics stimulated interest in postwar auto design. The 1920s saw the experimental teardrop-shaped "dream cars," mostly European such as the Tatra, embodying the latest airflow principles. Streamlined auto bodies became popular in the late 1920s and early 1930s when designers recommended more graceful lines for greater customer appeal in marketing. Headlights and fenders were the first items to be stylized. Designers and sheet mill technicians had to pool their skills to implement the designs.

Automotive Changes Steel

The feedback loops between automotive manufacturing and steel production methods were direct and immediate. Until the mid-1920s, wide sheet steel was produced in manual sheet mills. They consisted of two rolls placed on top of the other between which steel was passed and repassed until reduced to the desired thickness or gauge. As the sheet emerged from the rolls it was caught and pushed back over the top by a man at the back of the mill who was known as the catcher. The man at the front, the roller, took the steel with tongs and put it through the rolls again.[2]

Hand sheet mills first operated in the late 18th century and were little changed until introduction of the continuous sheet mill in the 1920s. In the hand mill, quality control was an art in the hands of the roller. Much more of the product lay in the mind and hand of the skilled worker than other steel products. Each pass of the steel required a manual adjustment of the rolls, which was dependent on the individual's judgment.

The improvement in the manufacture of steel sheet between 1920 and 1924 was not in the rolling end but in various finishing practices after the sheets were rolled. Demand for sheet grew through the 1920s as well as quality expectations. A number of heat-treating practices, such as annealing and pickling, were developed so that sheet met the specifications of the users.

Between 1920 and 1925, demand for light flat rolled products increased from 5.7 to 8.2 million tons in the United States. Production for the auto industry itself grew from 2.2 to 4.3 million tons. In addition, there was rising demand from producers of tin cans and other containers, plus makers of furniture and appliances.

The continuous strip mill and the cold reduction mill, where thicknesses were further reduced, were the most revolutionary technologies in the steel industry in the first half of the 20th century. Between 1924 and 1940, 15 million tons of new hot strip capacity was added to the American industry.[3]

There was a radical change in stamping requirements. Stamping is where a sheet of steel is punched by heavy dies into designed forms that become components further along in the manufacturing process. The contours were rounded instead of square. The body was assembled from a relatively few stampings, which meant a requirement for extremely large sheets with good drawing qualities, the metal's ability to be stretched.

In the 1930s there were few further technical improvements in the hot strip mill, but more importantly a huge expansion of capacity. The major new technical development in sheet mill operations was in cold reduction. After the hot mill, a long cold strip form was fed through a series of rolls at room temperature and reduced by 50–75% in thickness. This was crucial for the rapidly growing market for thin gauge sheet. It allowed production of sheet that was much thinner gauge than the capabilities of the hot strip mill. The initial appeal was to the container industry because the cold mill produced one-third less gauge variance between sheets than the hot mill and this was critical for the tin plate input to the high-speed, automated forming equipment now standard for can producers. Cold rolled sheet had much better surface qualities and more ductility so that sheets could be more easily formed into shapes that were required for automotive and appliance applications.

The Fisher Body division of GM introduced the all-steel roof or "turret top" in 1935. The innovation increased rigidity and strength, reduced vibrations, and allowed a lower center of gravity. One of the major barriers had been the unavailability of a stamping large enough to cover the whole top of the car in one single piece. Steel sheets were not available in suitable widths and machines did not exist to stamp large sections. The missing technology gap was the continuous wide strip mill.

Within the steel industry, the developments in auto design and manufacturing techniques required metallurgical qualities in sheet steel that would permit it to receive deep indentations without tearing or creating surface defects. The development of large presses also brought demand for larger sheets. The livelihoods of the steel companies, particularly in the early Depression period, depended on responding to the new demands of the auto industry. This often meant significant new investments in very tight times, just to stay in the game. By the mid-1930s, overall steel demand was down by 70% but light flat rolled steel accounted for 40%

of steel shipments. Continuous hot strip mills and cold reduction mills were often built in tandem.

Companies experimenting with new techniques in cold reduction for tinplate found that several of these processes also applied to producing wide sheet for the auto industry. Cold reduction reduced the thickness of the steel by 80%, increasing proportionately its length. In order to take up the slack in the product flow, increased speed was required at each stage (stand) in the process. In mills of the early 1930s, the strip left the first stand at a speed of less than 300 feet per minute and the last stand at more than 4000 feet per minute. In order that the speeds were appropriately synchronized at each stand, it was necessary for each to have its own power source. Electric motors and controls provided the power and regulation necessary for the operation. The cold reduction process gave the sheet a fine smooth surface and changed the grain structure of the metal in such a manner that after reheating or annealing the steel could withstand the stress of forming and drawing operations without breaking or tearing.

The new continuous mills required 50% less capital for equivalent capacity and employed 200 fewer men. Productivity in the hand mills was 0.1 tons per man hour. In the new continuous mills it was 1.3 tons per man hour. High volume output, with consistent quality, married mass sheet steel production to the emerging new consumer society.

The Modern Auto–Steel Interface

The core automotive production technologies—stamping, welding, painting, machining—are heavily determined by the choice of product technology with the result that there is considerable per-unit cost advantage in large, centralized manufacturing plants that capture economies of scale. In turn, this means that the industry is capital intensive and highly concentrated, requiring very large investments both in the manufacturing system and in each new model design. A typical modern car plant with a two- or three-model capacity of 350,000 cars per annum will require an investment of at least US$2.5 bn, while each new model platform will require US$1 bn. This system of production and consumption system emerged to meet particular conditions. In the early days of the motor industry, in the "craft" phase of car manufacturing, manufacturers could only deliver customized vehicles in low volume and at high price. As a result, there were many small companies that could be classed as vehicle manufacturers—typically several hundred in each of the major industrialized countries. The business model was one of designing while manufacturing, passing on the resultant high costs to affluent consumers. Ford changed the business model by introducing mass assembly of more standardized vehicles at low unit prices, but crucially the vehicle body technology (the rudimentary steel chassis) employed was preindustrial in character. Hence, in this instance, the redefinition of the business model not only preceded the redefinition of the product and process technology, but also provided the framework within which the all-steel body made economic sense.

Automotive industry economics are dominated by the choice of production technology for the car body. The auto OEMs have two core competencies in design and manufacturing: vehicle bodies and engines. Virtually all contemporary cars are constructed from 250 to 350 steel panels, pressed then welded to create a "unitary" body structure, which is then painted. This approach demands high R&D and production investments, in particular, for the tools and facilities needed to press, weld, and paint steel panels. The high capital cost is offset by fast production rates and high total production volumes, which generate economies of scale. While this per unit cost advantage cannot be denied, the all-steel body demands high volumes to achieve low per-unit cost.

For the global automotive industry, model platforms strategies were the panacea of the 1990s, the perfect way to combine economies of scale, globalization, multibranding, and rapid rates of new product introduction. Components from one vehicle are used on another in a more comprehensive and structured manner. The "platform" refers to the vehicle body structure: the floorpan, front and rear bulkheads, and key items such as the suspension. Moreover, the commonality of components between apparently dissimilar models can be extended to include the "running gear," that is, engine, transmission, brakes, and steering.[4]

By sharing components across several models, the vehicle manufacturers can achieve higher production volumes and drive down unit costs.

The production benefits of platform strategies essentially revolve around economies of scale and enhanced flexibility to switch between models. Hard steel tooling dies used to press steel panels into shape achieve maximum economies of scale at high output rates—typically 2 million units a year. Generic platform strategies allow this possibility to be exploited to the full. Moreover, with platforms the vehicle manufacturers can decide how to allocate tooling sets between manufacturing operations. In essence, this is a decision on whether multiple tools are used, or single tools, with parts shipped between plants. In turn, this means that the plants can be "loaded" more easily with longer production runs and fewer tool changes.

A different approach to platforms is notable: that embodied in the Fiat Multipla. This is essentially a modular steel space frame structure with nonstructural steel panels allowing key dimensions such as width and length to be adjusted easily. In addition, the technology has a breakeven point volume of only 40,000 vehicles per year.[5]

Advantages of the platform strategy include: manufacturers can offer "more car for the money" with higher levels of equipment, better materials, and so on. Differentiation between various brands in a multibrand structure is achieved in a variety of ways not just in terms of product appearance and performance, but also service, finance, and other packages.

The downside of platform strategies is the danger of reinforcing a tendency to oversupply the market. Platform strategies do not allow vehicle manufacturers to escape the need for high volumes to achieve low costs. They are a means to achieve such volumes. This presupposes the market's ability to absorb production. It is for this reason that the Fiat Multipla steel space frame approach is strategically important for steel, because it represents a product design and production philosophy for much lower volumes.

Platform strategies will reinforce two important trends in steel supply to the auto industry: high-volume globalization and flexible supply. Competition from other materials for body parts will continue.

In general terms, platform strategies are conducive for steel, retaining key applications in vehicle body structures as long as the steel industry can meet the wider demands being made by the automotive industry.

However, platforms will not result in elimination of competition from other materials.

The future of auto steel will be most impacted by the prospect of design and manufacture of cars moving beyond the steel unibody into space frame designs. This will open up a very different production and distribution model in the auto industry. It may facilitate a new round of materials competition in which plastic and aluminum components have a renewed opportunity as well as hybrid composite of steel and other materials in combination.[6]

Steel Clusters and Local Markets

For steel producers, there is a natural steel "cluster" of steel companies and their manufacturing customers who have to locate close by because the product itself, from BOF and EAF producers, have heavy transportation costs. The freight cost variable is the fundamental determinant for immediate cluster behavior in steel.

In this scenario, the steel mill is a hub and other businesses want to locate around it. There is also a segmentation of customers around the mills. For those using lower priced, commodity grades of steel, freight costs are the economic dividing line, but for those pursuing the higher priced, value-added grades, they need the steel mill's technology and engineering talent.

The steel-manufacturing cluster phenomenon is the site of traditional connections between mills and heavy manufacturers such as automotive and appliance fabricators often in a specific regional location like the Midwest.

Engineering and process improvement stories abound in the history of interaction between the integrated steel mills and the original equipment manufacturers (OEMs). Many relate to basic metallurgy, because so much of the final steel product attribute set is determined by the original metallurgical and processing parameters in the melt shop of the steel mill. This is where producer–user interaction has been closest.

Historical patterns of supply are evolving in accordance with changes in advanced manufacturing in general. The heart of the issue goes to the model of the auto industry supply chain, the lead customer for steel and the reference point for modern lean production.

We use hundreds of thousands of tons of flat-rolled steel.

The auto OEM Resale programme dominates. In most cases, OEMs purchase the steel, seeking bulk pricing from the steel mills and distribute the steel to the Tier 1 parts suppliers. From the mid-1990s this changed how we do our business with the steel industry.

There are two channels of steel supply: Resale is 65%, Non-ReSale is 35%.

Tier 1 Auto Supplier Executive

This new approach to manufacturing and supply has created different and not always welcome relations between management of the Tier 1 suppliers, the top auto parts producers and management of the steel mill.

On the ReSale, we get involved with logistics, quality, etc. Everything but the purchase transaction. The relationship with the Mills is good but not as good as if we had the whole transaction in our hands. He who pays the piper … The system dilutes our relationship and leverage with the Mills. On Non-Resale steel we have the service centres between us and the Mills.

Tier 1 Auto Supplier Executive

Nonetheless, as R&D responsibility has devolved from the auto OEMs to the Tier 1 suppliers, they feel the need and have the desire to establish more developmental relationships with the mills in the future.

For advanced parts manufacturers, technical interaction is the most important factor particularly for HSLA or Dual Phase steels. We work on very specific applications. There is no recourse if they are out of spec. The steels are prototyped from the design stage forward, which specifies certain grades of steel e.g., for certain stiffness characteristics. This is the importance of locally sourced steel.

We want to work directly with Mills on R&D, cost reductions and moving new grades to reduce costs.

Tier 1 Auto Supplier Executive

It seems that there is a substantial future for steel mills within their natural economic cluster, although there is a relationship rebuilding job to be done to work through the complex issues in the new manufacturing supply chains in order to be able to take advantage of it.

Postindustrial Steel: Is Pittsburgh Still There?

Virtually all public policy shops and policymakers at all levels of government now use economic clusters as their policy framework. The reference point for this major policy shift was the work of Michael Porter of Harvard Business School in his 1985 book *Competitive Advantage*. Porter's insight and argument were that competitive advantage did not flow to countries or to firms but to groups of firms (producers, customers, suppliers). The clear implication is that location matters in economics and economic policies.

There is a Steel Technology Cluster. It is composed of the steel producers and their suppliers of material and professional services (engineering, logistics).

To illustrate the importance of clusters, consider that contrary to many news reports and conventional wisdom, the steel story is not over in Pittsburgh. The conventional wisdom is that while the steel industry has lost its mills, the jobs have been replaced with health sector jobs at the University of Pittsburgh Medical Center.[7]

Although Pittsburgh lost most of its direct steelmaking *capacity* from the 1980s onward, it did not lose its steelmaking *expertise*.[8] The importance of this for jobs will be explained in some detail.

The Steel Technology Cluster is made up of firms that provide a diverse array of products and services as part of the supply chain of the steel industry. This supply chain can be divided into four main components:

1. Production equipment used by steel mills;
2. Engineering services that assist mills in the selection, design, and upgrading of that equipment;
3. Parts and supplies needed to keep that equipment operational; and
4. Raw material inputs to the production process.

Estimates of total employment in the Steel Technology Cluster around Pittsburgh are well over 12,000 people, with an average wage of $56,000. This represents a 50% increase over the average regional wage of $36,051 and a 10% increase over the current average wage for Iron and Steel Mills in the region of $51,000 in the past.

Contrary to the assumption about the disappearance of steel, the intermediate suppliers of goods and services to the steel industry have managed to not only survive the loss of steelmaking capacity in the region, but also transition successfully into an integral part of the global steel supply chain.

The development of the Steel Technology Cluster arose from the process of deverticalization of the steel industry. It has had two main effects on the role of intermediate suppliers. First, they have expanded their role in the supply chain to include services as well as products, such as the bundling of material handling with the supply of raw materials. Second, they have developed a network of relationships with each other in order to coordinate the supply of products and services to a global (rather than local) industry. Although geographic proximity to the customer is no longer as critical to the suppliers, geographic proximity to other suppliers has risen in importance.

Industrial clusters in a globalized economy do not subsist as islands in themselves. They exist in a series of nested scales. The Steel Technology Cluster is embedded in a larger Materials and Manufacturing cluster.

The economic performance of industrial clusters is traditionally measured by their relative export performance. On this basis, the steel industry has historically performed very well compared to other manufacturing industries.

Analytically and policy-wise, the economic performance of clusters has been strongly correlated with the phenomena of interfirm knowledge flows as well as the impact of high skilled and specialized local labor pools.

Recent academic work has drawn attention to the importance of "relational rents" as a more important factor than relative export success in examining the impact of industrial clusters. Rents as defined by economists are levels above competitive market levels of profitability. In the cluster context, these are gains beyond that reflected in traditional trade statistics.[9]

The classic case of relational rents in the academic literature is that of Toyota and its interaction with its suppliers. The network economics of the OEM–supplier agents generates significant rents that are then shared by the firms within the network.

Future research may examine the phenomenon of relational rents in the steel technology and material and manufacturing clusters as their most important economic impact.

New Global Steel Manufacturing

The steel-manufacturing cluster phenomenon is the site of traditional connections between mills and heavy manufacturers.

Historical patterns of supply are evolving in accordance with changes in advanced manufacturing in general.

Herrigel argues that manufacturing, in this case steel manufacturing, has a bright future, in the sense that it is irreplaceable given our consumption needs and the high recyclability of the material. Steel will be manufactured somewhere. However, it faces great challenges. Two factors will determine the viability of that future path.

In 10 years, we will not be making the classic distinction we conventionally make between goods production and services in the economy. Industrial production will only be seen as viable if it satisfies human needs and does not excessively impair the physical environment. We are all familiar with the exhortation for our industries to become more efficient in order to boost productivity, become more competitive, and sustain a high wage economy. All this is true.

It also means that manufacturing companies will increasingly resemble service companies instead of classic industrial commodity producers. As some steel companies used to say, they don't sell steel, they sell solutions. Understanding, leveraging, and taking advantage of the information content and design potential in the steel—its basis in advance metallurgy—will be key to how the steel industry manages its future. The critical success factor will be its fundamental capacities to innovate and its skills and human resource capabilities. Therefore, examination of innovation in the steel industry is a key theme in this book.

New Collaborative Supply Chains

A supply chain is a system of organizations, people, technology, activities, information, and resources involved in moving a product or service from supplier to customer. Supply chain activities transform natural resources, raw materials, and components into a finished product that is delivered to the end customer. In sophisticated supply chain systems, used products may reenter the supply chain at any point where residual value is recyclable. Supply chains link value chains.

Incorporating supply chain dynamics successfully leads to a new kind of competition in the global market where competition is no longer of the company versus company form but rather takes on a supply chain versus supply chain form.

There is often confusion over the terms supply chain and logistics. It is now generally accepted that the term Logistics applies to activities within one company/organization involving distribution of product whereas the term supply chain also encompasses manufacturing and procurement and therefore has a much broader focus as it involves multiple enterprises, including suppliers, manufacturers, and retailers, working together to meet a customer need for a product or service.

Value Chain

The value chain, also known as value chain analysis, is a concept from business management that was first described and popularized by Michael Porter.

A value chain is a chain of activities for a firm operating in a specific industry. The business unit is the appropriate level for construction of a value chain, not the divisional level or corporate level. Products pass through all activities of the chain in order, and at each activity the product gains some value. The chain of activities gives the products more added value than the sum of added values of all activities.

The value-chain concept has been extended beyond individual firms. It can apply to whole supply chains and distribution networks. The delivery of a mix of products and services to the end customer will mobilize different economic factors. The industry-wide synchronized interactions of those local value chains create an extended value chain, sometimes global in extent.

It is often commented that the majority of steels in a recently purchased automobile did not even exist 10 years ago. This stands in contrast to the public misperception that steel is an obsolete smokestack industry. On the contrary, as this book argues, innovation and new steels are a constant in the new steel industry. As one steel executive has said publicly, "This is not your grandfather's steel industry."

Innovation in steel is a complex process. It is sometimes driven by steel producers and sometimes the steel companies are pulled by their customers. Other times it comes from outside third-party sources. Some examples of each of these innovation paths are outlined in the following examples.

Auto Steel

This is the classic case of Customer Pull innovation. The quality and manufacturing process revolution symbolized, but not exclusively restricted to Toyota-ism, was a revolution not only in production processes but also for material inputs. It was the Transplant Japanese auto companies locating in North America during the 1970s that force-fed steel innovation into the operations of auto-oriented steel makers, particularly with galvanizing lines.

> *The tipping point and the driver [for dual phase steel] were the Japanese Transplants. The Japanese mills had developed it. The auto companies insisted on it, forced it on the domestic companies, otherwise they would have gone to foreign producers.*
>
> *Ex-Steel Company Executive*

The steel companies built on these innovations but they were dragged into the game by the transplants. It is not clear that North America would have had as innovative a modern steel industry in the 1980s and 1990s if it had not been for the Japanese auto companies. As a result, R&D expenditures in the past 20 years have been led by auto-related steels. This was the leading user market and that was where the best profit margins were found for integrated steel producers.

Manufacturing

In the coming years, producer-push innovation may provide an opportunity for steel producers to increase advanced manufacturing customers. Having put such tremendous resources into auto steels in the past decade, there may be major opportunities for applying the new metallurgical processes and products to non-auto manufacturing uses.

> *There are extremely poor technical capacities in manufacturing in terms of understanding and applying the new steels. The stampers let Toyota and the steel companies do all the work. Stampers just work on cost and yields from processing. There is no development. Their margins are so precarious.*
>
> *Steel Executive*

Many public policy shops are now advocating a rebuilding of manufacturing capacity as part of the post-Recession, and more sustainable economy. Nonautomotive application of auto steels could be a big contributor to this rejuvenated productivity and sustainability story.

Construction Steel

New applications of flat-rolled steel are a major emerging story. In this case, third parties outside the steel industry may be the key innovators. In the case of new coated and painted steels, it is the paint companies that are the lead innovators.

> *Construction is not like the auto side of things which always is talking about Grades, micro-structures, etc. In this case, it is the paint manufacturers who are the source of innovation. The paint suppliers push innovation at the steel producers, companies such as Valve Spar, PPG, Becker Coating.*
>
> *The paint guys call on us more than do the steel mills. We get incentives to utilize their new products then we push the steel companies.*
>
> *Manufacturing Executive*

This story is currently being played out in the mid-West. In manufacturing and construction, coatings are the key innovation. The steel producers supply the substrates.

Similarly, in the energy-rich western North America there is an exciting steel manufacturing story emerging around welding technology. Steel fabricators for energy projects are to some extent playing a similar role as the Japanese auto companies in the East. It is innovation in welding technology blended with new metallurgy that is changing the key determinant—welding—in steel fabrication. Again, like paints in construction, it is welding in fabrication that is driving new steel applications from outside the traditional industry.

As steel producers become more deeply integrated into supply chains, they become subject to new dynamics of production scheduling and inventory management risks. Steel supply chains have stretched out downstream as increasingly steel shipments have shifted from being primarily an interface between primary producers and original equipment manufacturers (OEMs), for example, U.S. Steel–GM, to where a majority of shipments, in the order of 70% are shipments from mills to service centers to manufacturing customers who themselves are embedded in complex supply chains either as Tier 1, Tier 2, Tier 3 suppliers to OEMs or manufacturer to wholesaler to retailer to final consumer. The complexity of steel operations management issues increases on a rising scale.

The financial risks are also increasing where traditional margin pressures and high fixed cost of mill operations of steel-producing companies are exacerbated by steel companies becoming the holders of the residual financial risks for the whole supply chain.

Does Steel Have a Growth Story?

For many observers, steel as an industry in North America will be flat or declining in the coming decade, depending on one's view of the auto industry and whether auto leads the downtrend in manufacturing as a whole.

There will always be an industry for high end products like auto at the 12–13 million vehicle level. But auto demand and therefore manufacturing demand will fall in the future. The high end with ULSAB will always be there but what of the rest?

*In the future North America will have 85–90 MTs of production and
130 MTS of consumption. The leaders will be high end auto plus oil
and gas pipelines*

Steel Consultant

In the conventional view, some regions may do somewhat better depending on demand for steel related to energy projects.

In a different scenario, there are two factors that might bend the flat/declining line in a more optimistic direction.

First, there is room for the development of nonautomotive applications for the new auto steels into other areas of manufacturing. This could mean an increase of 5% in steel demand. This will require a change in steel customers' attitudes away from simply regarding steel as a low-cost commodity. This would also mean manufacturing, retooling, and retraining required to apply the new steels.

Second, a set of policies might significantly facilitate the penetration of flat-rolled steels into residential construction and also other buildings and storage facilities. A more active building code and trades training policies would be critical. Optimists believe that the construction market for flat-rolled steel could be a major new growth opportunity beyond the existing market in rebar and beams and theoretically could result in a 20% market growth over time, equal to the auto share now.

*Construction is the elephant in the room. They could do Dual Phase
for lighter stronger applications in construction skins. If steel penetrates construction it will have a much greater impact on volumes
than further work in auto. Materials competition is the key strategic
issue. The steel companies don't see it.*

*Unless they can break into construction then 5–6 mills will go down
over the next five years.*

Ex-Steel Company Executive

If neither of these new growth dynamics occur, then the industry will be important but on a slow retreat into the future.

Steel Supply Chains and the Amplification of Business Cycles

The Bull Whip Effect (BWE) is the amplification of demand variation through a supply chain. It says that as variation in consumer demand increases, demand variation will increase at each subsequent upstream supply echelon, from retailers to wholesalers, manufacturers and their suppliers. The presence of a BWE causes increased uncertainty for managers, thus often leading to increased cost and reduced inventory management performance.

The original studies on the BWE in steel examined the dynamics between steel mills, service centers, and auto industry parts manufacturers.[10] An inherent systems risk was identified as demand variance increased back up the supply chain. The root of the problem was identified as the combination of the rolling schedules of the producing mills plus the discount policies of the service centers. Attempts have been made to manage this better through demand smoothing and in concrete terms, moving to vendor-managed inventories whereby the service centers take over this supply management function for their manufacturing customers. However, in the longer term this has turned the service centers effectively into working capital suppliers to manufacturers, a very different business model from what was originally intended.

Recent analysis of the 2007–2009 Recession in the United States has identified a good case for BWE factors present with the final impacts being most felt upstream by manufacturers and by implication suggesting new challenges for the steel supply chain.[11] The authors posited that increased demand variation due to the recession will be greatest from wholesalers to manufacturers, second most from retailers to wholesalers and least from consumers to wholesalers. Because inventory levels will also be subject to the same forces of variation, the authors further posited that inventories will vary most within manufacturers, second most with wholesalers, and least with retailers.

For 2007–2009, retailers appear to have attempted to buffer themselves from demand variability by smoothing orders and inventory. Conversely, wholesalers reacted most aggressively but seemed to have lost control of their inventories. Manufacturers, at the end of the chain, suffered the biggest impact as the amplified demand variability of manufacturers' inventories increased by 147%.

The BWE effect is not just of interest to operations managers and academics. The dots connect at the macroeconomic level.

As noted elsewhere, excluding food and energy, about 80% of consumer products have steel in them. The big picture connection is this: the retail revolution in supply chains has been more fully implemented in North America than in any other economic region. Inventory management and lean production norms are imbedded in management. It means that the new system was unusually sensitive to external shocks in the global economy. It is reasonable to posit that BWE in the retail-manufacturing customer base for North American steel producers was a major contributing factor to the 2008–2009 downturn in NAFTA steel being greater than in any other steel-producing region of the global steel industry.

Outside market forces have been and continue to be determinative of steel business cycles and foundational for the industry's future growth.

CHAPTER 9

Steel and Technological Change

Technology has been at the core of what the steel industry and its companies do. Major shifts in the business models for the industry have been strongly correlated with new production technologies at the furnace end and at the rolling and finishing ends of the business. Examples have been previously discussed in Chapters 3, 4, and 8. A high-level view suggests a series of periodic major changes such as the Open Hearth being replaced by the BOF then the EAFs and the continuous casters. For economists and modelers of change, the industry has been characterized by step-function changes in technology. This suggests a linear path of punctuated equilibrium—long periods of incremental changes then disruption by a qualitative shift, then incrementalism resumes. We may now be at one of those incremental periods, while awaiting a probably environmentally mandated change to blast furnace technology in the coming decade.

That the process will be continuous does not mean that it will be easy or straightforward, particularly when confronted by the traditional culture of steel companies.

The chapter begins with a discussion of the general trends of technological change in the steel industry from the perspective of evolutionary economics. We then look at process innovation in the steel industry. Then, we will look at examples of product innovation in steel from steel manufacturing users to consumer product. More generally, we will look at the new product development practices in global steel companies. Finally, we will look at the present and future positions of steel in the materials competition in key product markets, particularly with aluminum in automobiles and with concrete in construction.

Steel and Evolutionary Economics

For an industry as capital intensive as steel and characterized by periodic production technology shifts, concepts from evolutionary economics are critical for understanding the dynamics of industrial change. Concepts such as path dependency and lock-in have been major factors in how the industry and its companies evolve over time.

In terms of technology alone, steel is best seen as a case of punctuated equilibrium, that is, for extended periods the industry proceeds with incremental innovation. Then, along comes a transformative technology like the wide strip mill, BOF, continuous caster, and minimill that shifts the goal posts for the whole industry. The industry and companies adapt, or not, then the path of incrementalism resumes.

Path dependence explains how the set of decisions one faces for any given circumstance is limited by the decisions one has made in the past, even though previous conditions may no longer be present. It can refer either to outcomes at a single moment in time or to a long-run equilibria of a process. In common usage, the phrase implies either: that "history matters"—a broad concept, or that small differences are disproportionately amplified in later circumstances

However, the story of path dependency and lock-in are more nuanced than just the technical technology story by itself. For the most part, the corporation has been a black box for economists. This is a general issue in business history, not just a steel problem.

Recent research has lifted the lid to open the black box of organizations and to view organizations as competing on the basis of their routines that are built up over time, where routines are seen as organizational competences that are more than just the skills of individuals. This perspective assumes that the success of firms is determined by routines built up in the past, that is, path dependence; and, views price signals as the main determinant of locational behavior. This evolutionary perspective has translated into a focus on groups of firms that have developed durable routines and habits, with the market environment operating as the mechanism of selection.[1] At the same time, path dependence need not lead to or involve lock-in, or indeed lead to any form of equilibrium or stable state or trajectory.

If we fully elaborate this perspective, then an industry like steel, not-withstanding the global pipelines of technology, will always function in specific, unique "steel economic spaces." There will always be a German steel industry, an American, an emerging Chinese steel industry. This is the basis of steel "clusters" distributed around the world. Economic geographers have consistently demonstrated the local bases of the mechanisms underpinning path dependence—for example, increasing returns and external/network economies—considering it to be "a process or effect that is locally contingent and locally emergent, and hence to a large extent "place dependent." In summary, for an industry like steel and perhaps for all of manufacturing, place matters in determining the nature and trajectory of evolution of the economic system.

At the nonlocal level, steel companies operate as nodes in a global production network (GPN). The boundaries of the firm are in fact fuzzy. Firms are composed of internal, company networks embedded in external social networks. As the geographic extent and complexity of company operations increase, the nature of their local embeddedness becomes far more complex. Local social dynamics, politics, and culture are important and have direct impacts on managerial decision making.

What does this Mean?

The emergence of international regulatory bodies like the World Trade Organization (WTO) and the International Standards Organization (ISO) shape the geography of different industries. For example, the multifibre agreement and its abolition in 2005 dramatically impacted the shape and distribution of the global textile industry. Operations of global companies are actually enhanced by things like ISO through the introduction of codifiable standards. At the same time, nation states are still critical. The media often say that nation states are irrelevant in a globalized economy but look at the Chinese example and the completely different geography of automobile production in China versus Eastern Europe. Finally, macroregional economic arrangements like NAFTA are critical. They change the surface on which industries and companies play. Where regional integration occurs it attracts inward foreign investment. The evolution of the NAFTA steel market as a clear example.

Taken together, all these factors affect technology development. The broader social and political environments will affect the directions and trajectory of technological change within the steel industry.

Steel Company Technology Development

Technical innovation has always been important in steel, as in all capital-intensive industries. However, it has taken on heightened importance in the past decade while changing its character and focus.

The emergence of the modern industrial corporation was closely linked to the development of technology in their internal corporate laboratories. The pioneers were German companies like Siemens in the late 19th century and followed in the early part of the 20th century by American corporations like DuPont. By mid-century, all major industrial corporations had developed large, specialized laboratories for product development.

In steel, this development came later. Early steel innovation was led by engineers focused on process innovation. It was in the late 1960s and 1970s that steel companies established separate R&D centers. The leader was U.S. Steel who once employed as many research scientists and engineers as all the rest of the steel companies combined. They were then followed by other leading steel companies. Research engineers became the technical leaders for their domestic steel industries in the postwar period. It was essentially an internal or indigenous model of innovation.

In the 1990s, the world of steel innovation changed. The major companies cutback or cut out their research and development facilities. They believed they were fighting for their very existence and could not afford such luxuries. If necessary, they believed that they could always license the latest technology from others.

A limited set of European, Japanese, and Korean steel companies, along with leading equipment vendors, became the technology leaders in global steel. Deep metallurgical engineering became the specialty of a limited number of global players like Nippon Steel and NKK in Japan and Usinor in Europe. The other steel producers increasingly depended on technology transfer and licensing or "traded knowledge," for example, depending heavily on NKK in Japan for steelmaking technology and Usinor in France for automotive applications.[2] Even today, the business press confirms this development with regular reporting on the latest technology licensing agreement between Japanese, Korean, and German steel technology firms for the building of the new Chinese steel plants.

This was how the companies dealt with production technology innovation. The second stream of technical innovation was on the product side.

For commercial application development, steel companies increasingly adopted the metaphor of software platforms and applications. They would develop specialized, local product market applications based on underlying languages (metallurgical technologies) that they licensed from others. The most profitable integrated steel companies in North America developed this strategy in the 1990s.

The following are three cases of technological change in production process technology, which illustrate the step and incremental character of steel technological change. Technological change in steel has been lumpy. They also suggest, particularly in steel, the simple distinction between process and product innovation breaks down. Much on the product innovation actually takes place at the "hot end" of the mill.

Technology Cases: The Open Hearth and Basic Oxygen Furnace

The U.S. steel industry built 22 new blast furnaces for ironmaking during World War II. Half were funded by the private steel industry and the other half by the Defense Plant Corporation. This amounted to approximately 10% of the capacity of the industry but included the largest and most efficient facilities where blast furnace man hours per ton had fallen from 1.1 to 0.56. There were few technical improvements in the furnace itself. However, improvements in refractory brick tripled the time between required relining and the flow of material to and from the furnace was significantly improved. The greater capacity expansion took place in the Open Hearth steelmaking shops. Instrumentation, which had been introduced to a limited extent in the 1920s, was extended. The improvements included instruments and controls for furnace pressure, temperature, fuel volume, and air volume, as well as devices to determine the amount of carbon in the heat of steel while it was being melted. These controls all contributed to the improvement of the quality of the steel and reduced fuel consumption.

During the war, the new developments in light flat-rolled products took a back seat to heavy plate and structural steels. Plate for ship construction led the way and most of the hot strip capacity was used for this purpose.

The Basic Oxygen Furnace (BOF) was introduced in the early 1950s as a substitute for the Open Hearth in steelmaking. It reduced heat times from 6 h to 40 min, at about half the capital cost. It was the classic Step Function improvement in technology. However, as described in Chapter 3, the US industry for cultural and political reasons, lagged its competitors in adopting it.

To compete with offshore producers, the U.S. steel industry had to improve both yields and quality. Until around 1980, yield numbers, that is, the ratio of final steel to raw steel produced produced, hovered around 72%. BOF and continuous caster producers were getting 80–90%. It required the production of a lot more raw steel to meet shipments than is the case today. The presence of semifinished steel imports (slabs, billets) have increased in recent years is a complication in the numbers but the overall ratio of raw steel to shipments is now close to 100%.

In the 1980s, the minimills were looking to move up the quality chain and produce more profitable value-added products, while the integrated mills collaborated with the Japanese steelmakers to install slab casting machines and maintain their share of the domestic auto market. The installation of continuous casters across the industry increased US raw steel yields dramatically.

From Continuous Casting to Continuous Process

The new steel technologies of the 1970s and 1980s, continuous casting machines, ladle refining, and vacuum degassing not only reduced significantly the number of man hours to produced steel but also led the way into the quality revolution and entire new product markets.

These technologies gave steelmakers the ability to fine tune the chemistry in ways that enhance characteristics like surface quality. Whereas the manipulation of steel chemistry had taken place in huge BOF furnaces previously, hot steel from the BOF is now directly fed to a ladle where temperature and chemistry can be controlled much more precisely. Further refinement in vacuum degassers has had much the same effect. By moving hot steel directly from the BOF furnace to an environment where that material's properties could be carefully controlled, immense quality improvements were realized and new steel products were developed.

Traditional steelmaking constituted a series of batch steps, for instance with inventories of solid steel between the BOF and the down-stream rolling operations. Molten steel would come out of the BOF to be poured into ingot molds, where it stood until it solidified. Then it would be stored to await the next stage. In addition to the time, cost, and energy required reheating the steel, the steel had to be committed to a semifinished shape—billets or blooms for long products and slabs for flat products. Rolling operations followed; each requiring reheating.

The advantage of the traditional process, relying on ingots, was related to the fact that integrated mills produced a wide variety of products. With an inventory of ingots, material could be fed to any of a number of lines producing different semi-finished shapes. If there were orders for rebar then the ingots were fed to a billet mill. Steelmaking and rolling operations could also use the ingot inventory as a buffer if there were disruptions or coordination problems. The BOF could continue producing if there were problems in rolling, and rolling could continue to operate when there were problems in steelmaking.

Continuous casting changed this process radically. Every caster is built for one and only one semifinished shape. It is dedicated to the product market that has been chosen. Once that strategic choice has been made, steel coming out of the BOF is never allowed to solidify before the casting operation. It goes directly from molten steel out of the furnace into one of the three basic semifinished shapes by passing through the continuous caster.

Steel Production Technology: The Future

The massive investments in new technology by steel companies in 1980s did not bring equal results. Studies of the combination of efficiency and technology implementation indicate very loose correlation between investment and results. Firms with comparable levels of investment could be 50% different in their results.

The differential had to do with the culture and presence of all the factors contributing to high-performance workplaces. Integrated steel plants, in particular, are composed of a series of processes and all the processes must be compatible and coordinated in order to achieve maximum efficiency and quality of product. Generally similar results were found for the much simpler processes of EAF producers.

With ingots gone and inventories eliminated, steelmaking became a continuous process. It is no coincidence that steelmakers invested heavily in ladle furnaces and vacuum degassers at the same time that they were investing heavily in continuous casters. The casters made it necessary to learn to work with hot steel and keep it flowing. It is a small matter to extend that control to intermediate steps at a ladle or vacuum degasser for adjustments in chemistry before proceeding to the casting machine. Success in coordination breeds further success.

Before the advent of ladle metallurgy fine-tuning the chemistry of steel really meant guesswork in adding needed alloys. Now, a steelworker in a control room high above the furnace keeps track of half a dozen control screens, checking tolerances in the chemistry and making changes as required. It illustrates the point that human capital has to be aligned in conformity and with equal dedication to capital equipment investment.

There has been a two-pronged process of decentralization of authority downward onto the shop floor at the same time as traditional layers of management above shop floor workers were eliminated. U.S. Steel, for instance, eliminated three layers of management, just within its operating mills. For instance, works managers, who were primarily responsible for daily operating decisions, have seen these responsibilities move to foremen and operators on the shop floor. Works managers now primarily deal with quality assurance and production planning.

Studies in the mid-1990s emphasized two trends for the future.

First, there will be a convergence at the steelmaking production technology end to blur the distinction between integrated and minimill producers. Both would be using sophisticated additions to their traditional primary charges of iron ore and scrap, allowing that there would still be differences in melt chemistry. However, both would use thin slab casting to feed generally similar hot rolled flat products and cold-rolled/galva-nized flat products for the same markets in the future.

Second, organizational development would focus on fewer layers of supervision and bureaucracy, along with greater employee involvement, responsibility and training, ultimately relying on self-directed work teams on the shop floor. In this respect, the minimill style of work organization and decentralized management was expected to become the norm for the entire industry. This is discussed further in Chapter 11.

Steel Product and Process Innovation

Research on the steel sector has found that most of its innovation has been on the process side. This is because at the end of the day steel for most customers is still a commodity where the number one competitive advantage is price. While coatings and light-weight steel are important to markets such as automotive, in the end you sell steel because of a competitive price and an acceptable level of quality. The reality of steel as a commodity business dictates that research tends to focus on ways to make steel cheaper and to process it faster so as to be able to outbid competitors. Steel's status as a commodity forces innovation on the process side.

This focus on process side innovation is reinforced by the relationship with customers. Customers from automotive to construction also tended to view steel as a simple commodity purchase and this limits the steel sector's ability to develop new products. As the quote below illustrates, many customers are not willing to pay a price premium or be reliant on only one producer and this provides a disincentive to mills who wish to create new steels.

Maybe not so in the manufacturing processes but in steel products a patented steel product at least in automotive is not necessarily that great a thing. And that's what I found out in this product that we developed. Because we were quite excited about it and went out to the automotive industry and we're telling them about this…it was a win/win because it was a lower cost product to make but it had enhanced properties. So what better could you ask for. But the way the automotive industry works is they don't want a single source of buyer of any steel product because they want multiple suppliers of the *product that they can feed-off against each other to lower the price. 'But if it's only one company that can supply it that's great but unless I have two people that I can put against each other to lower the cost I'm really not that interested in it.' So until our competition catch up with us on that particular product it's of really limited value. So it was an eye-opening experience for me. Now on the process side that's probably not true. If I have the ability to make steel for $20.00 a ton cheaper then that is a huge advantage.*

Steel Executive

Innovation within the steel industry has been traditionally led by engineers. They are the dominant vectors of technological change. But, engineering functions and labor markets are changing as discussed in Chapter 11. What is observed is a "thickening" of the engineering labor market. The boundaries and hierarchies between engineers, technologists, and technicians are becoming more overlapping and blurring. Two of the relevant implications are: The total cost of R&D may be reduced, as a result, to the advantage of local firms, and, increasingly technologists have taken over the lead role on the shop floor in process improvement engineering.

The standard picture of engineers for most of business history has be one of technical employees who undertook in-house industrial research to transform basic scientific research into activities to enhance the market position and profitability of firms, thereby increasing the aggregate rate of growth in the economy. A more nuanced view stresses their role in standardization as a means to increase the diffusion of technical knowledge and innovation by reducing the boundaries of the firm and increasing interdependence.[3] This enabled industrial processes to facilitate the flow of standardized processes across the boundaries created by business organization. Professional engineering association emerged in the early 20th century, such as the American Society for Testing Materials (ASTM), the Society of Automotive Engineers (SAE), and the American Railway Engineers Association (AERA). Industrial standardization was one of the major activities of industrial and research laboratories from World War I to World War II. At the turn of the century, steel producers made concerted efforts to standardize certain products because they were increasingly required to produce a proliferation of similar products according to specifications drawn up by different customers. The producers' motives were to reduce costs and increase exports. Specifications were typically detailed lists of physical and chemical criteria for products or materials to be purchased, sometimes including criteria for testing and sampling the material as well.

Innovation for Steel Consumer Products

The linkage between key steel production technology developments and downstream products has been noted. The following takes a number of steel-consuming industries as examples. The story of steel and the auto industry has been described above, particularly the tipping point represented by invention of the wide strip mill. But the same technology ushered in a wide range of innovations for other consumer products.

Steel Cans

The container industry also benefited from developments in cold reduction techniques, that is, taking steel sheets and processing them into thinner and longer dimensions by passing them through successive "stands" or sets of rollers that essentially squished the sheet dimensions downwards and outwards. In addition, they could have tin and other coatings applied. In the 1930s, the tinplate market shifted from demand for plain, ordinary tinplate to a product carefully made in accordance with elaborate specifications according to the corrosiveness of the product to be packed and from the probable physical abuse of the containers. At one end of the spectrum, the softer end, there were tinplate specifications for friction tops for cans of fish and bodies for cans of vegetables. At the stiffer end were specifications for beer and soft drink cans where pressure was strongest.

Refinement and commercial production of the beer can was also a major preoccupation of steel producers and beer companies. The great pressure built up during sterilization caused standard food cans to buckle at the ends as well as straining the soldered side seam causing leaks. Improved forming and soldering practices solved the weak side seam problem. Adding phosphorous at the steelmaking stage solved the buckling ends problem.

The more challenging problem was the development of a more suitable lining for the beer can. Tin in the plate combined with a protein in the beer to produce a cloudy mess. Normal enamel did not work but a compound resin substrate with several coats of enamel solved the problem. To keep consumers from drenching themselves, the ubiquitous tab opener was developed at the same time. The beer can was brought to market in 1935. In its first year, notwithstanding the ongoing Depression, 160 million beer cans were sold. By 1953, sales were over six billion cans.

A greater challenge for steel can manufacturers was access to sufficient quantities of tin, in the late 1930s and even more so during the war when sources in Asia were cut off. Electro-tinning operations consisted of running the steel strip through an electrolytic bath in which large electrified tin anodes were submerged at set distances. The first commercial facility started production in 1937. Refinements during the war resulted in satisfactory results with one-fifth the required tin from traditional methods. Specifications for low tin and tin-less cans were developed during the war. Eventually tinplate substitutes and organic coatings were developed to lessen dependence on tin for some applications.

New Steel Appliances: Refrigerators, Stoves, and Washing Machines

Wide sheet steel also found increasing applications in appliances. A steel cabinet for a refrigerator could be made from a single sheet in a wide variety of shapes. The one-piece cabinet was more attractive and it eliminated problems of decaying insulation. Enameling operations for the new cabinet refrigerators were challenges not solved till 1933 with new paint enamels. Prices fell from $310 in 1927 to $130 in 1940.

One-piece, steel-bodied stoves also appeared in the 1930s. This allowed improvements in design, the range taking on a cabinet appearance with smooth lines and surfaces. It was the culmination of developments in stamping, enameling, and low carbon steel.

By the late 1920s the washing machine had basically evolved to its modern form. The coating problem being solved by Maytag in 1937.

Postindustrial Steel: The Rise of Digital Manufacturing

Higher-value-added products have been identified as a survival strategy for the steel industry. With declines in automotive and general manufacturing, greater steel penetration in the construction sector is seen to be a key growth story for the future.[4] The new-technology competition in construction is led by the fabricators rather than the steel-producing companies themselves.

Critical shifts are underway in the skills composition of the fabricating companies. The traditional knowledge base of craftsman welders and fitters are being displaced by CADCAM and BIM software systems. However, at the heart of design–fabrication interface, the critical visualization skills of the traditional craftsmen as embodied in the Fitters and Detailers is still required in order to formulate how a structure goes together. Steel manufacturers have increasingly teamed with graphics design studios to help produce the next generation of digital fabrication. They call it digital manufacturing. By combining traditional craft skills with graphic design skills, steel fabricators are seeking to position themselves as owning and controlling the DNA of building structures for the future.

The technical innovation is the conjoining of a traditional steel fabricator's competencies with new media and design talent from the "new economy." A prime motivator is that a steel fabricator faces a skills crisis. Its traditional base of welder-fitters is rapidly aging, with a special challenge in replacing the key tradesmen who have the capacity for visualization of how steel structures are assembled as the building goes up. The older work force is retiring and there are no replacements in the local economy. In an actual case, the CEO identified the HR issue as the need to recruit design talent. The key recruit was a steel sculptor who had a background in design, new media, and welding and ran his own graphics design company. This itself is an interesting transfer of knowledge and skill from the arts sector to manufacturing. The design entrepreneur himself had recruited industrial design talent from the auto industry along with designers with experience in household goods. To this were added art college and community college new media grads. Interestingly, they do not seek architectural students. The fabricator bought the design house, as now have several of their competitors. This esoteric combination of talents has become the technical anchor for an emerging $500 million digital manufacturing and fabrication company.

New Steel Product Development Practices

The generic new product development process for steel companies can be summarized as follows and is standardized as a methodology by local mills across global operations but with significant room for variation depending on regional markets.

- New steel products generally have some enhanced feature when compared with existing products. Examples include improved surface quality, increased strength, improved ease of forming, and improved corrosion resistance.
- The process employs the use of cross-functional teams with representation from all functional areas that are involved with manufacturing, marketing, and sale of the new product.
- The process employs a "Staged Gate" process. In other words, the development is broken down into a series of steps or Stages with decision points or Gates between each step.
- In preparation for Gate meetings a Gate report is prepared. At the Gate Meeting a presentation is given to the Gatekeepers summarizing the report.
- Gatekeepers include local management representatives from all areas including manufacturing, commercial, financial, etc.
- Support by Gatekeepers is required to move the product along to the next Stage of development.
- Decision to promote a product to the next Stage of development is based on a predetermined set of criteria.
- Development of products is coordinated by a Product Strategy Board (PSB), which includes local representation. The coordination of product development between plants is done in order to maximize efficiency and minimize duplication of effort.

Within this procedure, some examples of types of current product developments include the following:

Type of steel product	Product enhancements
Automotive Advanced High Strength Steels (AHSS) Examples: Dual Phase, TRIP, Hot Stamped, Martensitic, Stretch Flangeable, etc.	Higher strength while maintaining formability Improved crash energy management Ability to reduce thickness/weight of parts
New Steel Coatings	More environmentally friendly coatings Coatings providing improved corrosion performance

Steel companies also have gotten much more involved in the manufacturing processes and even cost management efforts of their customers.

An example is the appliances industry. The steel companies engage in a "teardown process." It is a technical approach to value creation for the customer. They systematically and rigorously disassemble the customer's appliance product. They brainstorm on cost savings, evaluate the manufacturing process (stamping, fabrication, assembly), design, material utilization, and quality. In recent years, this has resulted in tens of millions of dollars of savings that flow to the manufacturing customer.

The steel companies also engage in co-engineering support. Much of this involves the use of the state-of-the-art predictive tools to assist their customers in product development and improved material utilization.

Suppliers are also involved in disseminating new technology. Because their goal is to sell new equipment to their customers, suppliers are always letting their customers know about the latest advances in technology. Many firms rely on this dynamic to ensure that they are using the best equipment and materials available for their processes.

Conferences are important not only for marketing to customers but also for keeping up-to-date on what other competitors are doing, and what new markets they are targeting. Well-established, specialized industries, such as the steel industry, are more reliant on trade organizations for information. Organizations like the AIST often have their own political systems within, meaning that business contacts are often heavily linked to involvement in these organizations. The AIST has local chapters and meets regularly to discuss matters relating to the industry, including presentations by member companies on their newest product or service.

In terms of innovation theory, product innovation in the steel industry is no longer the internalized, top-down process it once was entirely within the clear boundaries of the firm. It now takes place in a broad "loosely coupled" network of relationships and interaction among facilities, customers, and even educational institutions, where the local is very important but so also are linkages between local and nonlocal actors.

The Future of Materials Competition

How steel performs in the New Economy in the future very much turns on how well it competes with other materials, partially for energy and environmental reasons. Its future is also dependent in part on where it is placed in the merging of materials and manufacturing.

> *Everyone knows steel vs. aluminum and the huge resources in people, R&D and marketing and PR have gone into it. UltraLight Steel Auto Body (ULSAB) was its poster child. If a similar effort was made in construction in contending with cement, brick etc., then it would have a big payoff for steel volume. If steel penetrates construction in these ways it has a much greater impact on volumes than further work in auto.*

> *The cement and wood guys are around lobbying on the Building Codes all the time. A new steel requires new codes and specs, all different. However it is not tougher than trying to get all the different car companies to agree on the use of new steel for a strategic frame part or something.*

> *Steel Executive*

The competition between materials is well illustrated by the contrasting auto and construction steel cases.

For the past 30 years, there has been intense competition among steel, plastic, and aluminum for their respective places in the future of the automobile. Projections of future aluminum or plastic body cars have been more the stuff of science fiction than what you can observe in parking lots or car sales lots. The steel industry has mounted a vigorous forward-looking technology development vision around the (ULSAB) Consortium. The case for steel is that the new products are lighter and stronger, with better surface qualities and much better energy efficiency and recycling records than either aluminum or plastic. This is a true and an under-appreciated story.

It is also the case that there is 85 years of experience with the steel unibody in automotive assembly, tooling and skills, from assembly workers to auto engineers, all embedded in manufacturing practices. The competing materials have huge learning and retooling challenges to overcome if they are to ever be really competitive with auto steels.

Fundamentally, the material competition in auto is around the respective metallurgical properties of the product.

It is a very different but equally interesting story in the new construction steels. Here steel competes against wood, concrete, asphalt, and more. The competitive barriers and challenges in construction are less issues of metallurgy than other factors external to steel or to manufacturing. If the new steels are to really penetrate and take off in construction then they have to confront the barriers that are embedded in building codes, construction regulations, and building trades' certification and training.

The challenges to new steels in construction are not metallurgy. They are regulatory reform and human resource policies. This is a very difficult and different challenge than in automotive. However, if quantitative growth in North American steel demand in the future critically depends on construction, then this will have to find a place in the public policy agenda of the industry.

CHAPTER 10

Regulation

Two of the most important regulatory factors in the history of the steel industry have already been covered: labor market regulation and trade disputes in Chapters 6 and 7. They are so critical and nuanced, they warranted special treatment.

We now turn to competition and environmental policy.

Because of its history of oligopolistic pricing, inevitably the steel industry came up against antitrust issues. The Pittsburgh Price system was dismantled just at the time that rising import pressures shifted the regulatory battle to Trade Law. Those were the dominant regulatory issues of the 1960s to 1980s.

However, regulation should not just be seen as a compliance issue in the perpetual debate between the role of free markets and public policy. Regulation can also play a positive role in facilitating product innovation and market development. An example is the SAE (Society of Automotive Engineers) product standards that originated in the early 20th-century automotive industry but have become the global standards for steel and manufacturing.

More recently, environmental regulation has preoccupied the industry and driven many of its capital allocation and investment choices. The rise of environmental concerns put steel at the center of issues involving water and air quality and recycling.

Steel Pricing and Antitrust: The Pittsburgh Pricing System

The role of competition, price setting, and commercial policy was the major single regulatory issue in the era of Big Steel in the first half of the 20th century.

The effect of U.S. Steel dominance in the industry was its imposition of the Pittsburgh Plus pricing system on the entire country. This system dictated that all steel prices be based on the costs of production and transportation from Pittsburgh, no matter where the steel was originally produced. This allowed producers based in Pittsburgh to compete with local producers all around the country, since these producers were unable to undersell steel made in markets that U.S. Steel dominated. Although its origins are obscure, Pittsburgh Plus was firmly in place by 1901 and U.S. Steel championed its continued existence. Despite losing a suit by the Federal Trade Commission in 1924, U.S. Steel fought to keep the Pittsburgh Plus system in place in a modified form until it lost a US Supreme Court decision on the matter in 1948.

The system was in large part founded on regional economics. The emergence of a large iron and steel industry in the Chicago region during the 19th century was a function of entrepreneurial effort and geographical advantage. Mills could obtain raw materials from the vast iron ore deposits in the Lake Superior region relatively cheaply and easily. Because most of the iron ore used by the American steel industry during its rise was mined in Minnesota and Michigan, mills located along the Great Lakes were well positioned to enjoy lower costs than their competitors elsewhere. They used the "Pittsburgh Plus" pricing system to protect Pennsylvania mills from competition in other regional markets like the Midwest.

However, the advantages to U.S. Steel itself were only temporary. In the longer term it is not clear that the company itself was well served by the monopolistic pricing regime.

U.S. Steel was founded with the hope that its size would lead to economic benefits in the form of market power, operating, and network efficiencies. While some cost savings were achieved, and some of the superior management of the Carnegie companies did transfer to the new entity, in the grand scheme of things, size appears to have acted as a drag on the new company rather than as an advantage. Warren[1] hints at some of this tension when he notes that those "who had worked at Carnegie found it difficult to work in reasonable amity with rival companies rather than competing ruthlessly with them as in the past." In the face of attempts to manage prices and exercise leadership, U.S. Steel saw its many small rivals eat away at its markets. Between 1901 and 1927, U.S. Steel's market share in raw steel dropped from 65.7% to 41.1%. This period of U.S. Steel's relative competitive latency allowed for the growth of companies such as Bethlehem Steel, which by 1903 was run by U.S. Steel "defector" Charles M. Schwab. U.S. Steel's price leadership strategies in the early 20th century may have led to a high return on sales, but the company's dollar sales were essentially stagnant despite significant additions to capacity and production.

By 1936, the stagnation at U.S. Steel was such that Fortune magazine "recalled that the Corporation's policy had once been summarized as "No inventions: no innovations" and Charles M. Schwab reported that the chairman of U.S. Steel admitted to him that the Corporation, in fact, had missed every "new thing" in steel. Under Carnegie's reign, Pittsburgh-based steel facilities had competed successfully against growing location-based advantages of other regions by relying on innovation, efficiency, and superior management. U.S. Steel under Gary did participate in the geographic dispersion of American steelmaking but its "Pittsburgh Plus" pricing led to a hobbling of the corporation's growth into new markets and eventually shrank the region in which Pittsburgh-area steel was competitive. Even in its infancy, U.S. Steel was an illustration of inertia and captivation by sunk-cost investments.

The era of price fixing for Big Steel came to an end. The Anti-Trust actions eliminated the system just as the competitive technologies of foreign producers were being introduced and, as described in Chapter 6, collective bargaining was becoming the central price-setting mechanism in the industry.

The Role of SAE Standards

The impact of regulation is most often described in negative terms for its alleged adverse effects on management, costs, and free markets. However, regulation can also play a major role in expanding markets through development of industry standards. A clear example comes from steel and automotive standards.

SAE (Society of Automotive Engineers) standards have become pervasive metrics not only for steel inputs into automobile production but also for general manufacturing usage throughout the global steel industry. The setting of industrial standards for steel materials goes back to the earliest days of the auto industry and the precursor of the SAE, the Materials Branch of the Association of Licensed Automobile Manufacturers (ALAM) in the first decade of the 20th century.

In 1910, engineers in the young American automobile industry initiated an extensive program of intercompany technical standards dealing with dimensions of parts and accessories, specifications for purchased materials particularly steel and engineering practices such as the design of screws.[2] The earliest steel specifications were set by the Mechanical Branch of the ALAM between 1905 and 1909. The Association comprised the numerous small assemblers, outside of Ford and GM, who each had their own unique parts designs and found it almost impossible to obtain parts elsewhere if their regular parts suppliers failed. In 1910, the SAE was formed and its engineer members eventually became the leaders in setting technical standards for the industry.

The Standards Committee of the SAE was divided into divisions, each of which developed standards for specific groups of parts or materials. For instance, the standards for steel tubing reduced varieties from 1,100 to 150 by 1911.[3] The most stringent material specifications were for iron and steel alloys. Previously steels had been sold by brand name or by each manufacturer's own specifications. It was estimated that through the standards, the steel costs for manufacturers were reduced by 20%.

At first, it was the smaller auto companies that created, followed, and acknowledged the value of the standards. However, by 1915 GM took an active interest in the SAE standards, particularly those for steel, and one of its metallurgists became Chair of the Standards Committee. GM was less concerned with detailed specifications for individual parts, because like Ford, it had imposed its own internal standards for its internal parts divisions. This reflected the fact that external purchase of parts by auto companies had fallen from 55% to 26% during World War I. Large automobile manufacturers did not purchase many small parts, but they still depended on materials and equipment from other industries. Here standard specifications continued to be immensely valuable.

By the mid-1920s, GM engineers were the most prominent in the SAE. Five of the Standards Committee's division chairmen or vice-chairs were from GM and they were on 16 of the 21 committees dealing with motor vehicles. The other major car companies, other than Ford, were also active participants. By the early 1930s, the SAE Handbook paid most attention to standards relating to systematized interindustry purchasing. The detailed standards for automobile parts, so important to the parts-purchasing automobile builders of earlier years, had disappeared.

The automotive industry by the 1920s primarily used SAE standards for purchase specifications. Interestingly, the bitterest opposition to purchasing standards came from suppliers of steel. In the early years, many of them deemed SAE chemical and dimensional standards for steel as an inappropriate intrusion. Producers of spring steel particularly objected to being told by what specific method they were to produce their product. As the volume of steel purchasing by the auto industry increased in the 1920s, steel producers came into compliance and cooperated with the use of SAE standards as a way to systematize the purchasing process.

Steel, the Environment, and the EPA

For current steel management, labor market regulation has been displaced by environmental regulation as the thing that keeps them awake at night.

Technical experts in steel believe that over the next decade the determinative variable in future technology trends within the steel industry will come from outside. They will be driven by environmental and energy policies. Ironically, at the same time, steel has more than met the much disputed Kyoto GHG standards over the past decade.

The steel industry has identified climate change as a major challenge for more than two decades. Long before the findings of the Intergovernmental Panel on Climate Change (IPCC) 2007, major steel producers recognized that solutions were needed to tackle CO_2 emissions. They have been highly proactive in improving energy use and reducing greenhouse gas emissions and are now operating close to the limits of the existing steel production technologies.

Even the best steel mills are now limited by the laws of thermodynamics in how much they can still improve their energy efficiency. With most major energy savings already achieved, further large reductions in CO_2 emissions are not possible using present technologies. The kind of further reductions being called for by governments and international bodies require the invention and implementation of radical new production technologies.

A set of breakthrough technologies is needed; the kind of paradigm shift in industrial production that can change the way steelmakers around the world operate.

Various research programs have already identified more than 100 new technologies, and classified them in terms of the CO_2 reduction they could achieve. Some technologies are ready to use but would deliver only a small reduction in CO_2 emissions. The more ambitious projects in terms of CO_2 reduction are now going through various steps of scaling up from lab to commercial reality.

The coal-based ironmaking technologies associated with carbon capture are the most likely candidates for early viability. Hydrogen and electrolysis are further into the future, as these technologies will require deeper reengineering of steel production and the development of new processes from first principles. Biomass solutions are probably in the intermediate future. In the even longer term, new avenues of research are likely to emerge. These include the integration of steelmaking with solar power generation, with new energy technologies and with new, fourth- or even fifth-generation nuclear power plants. Such solutions are not yet part of the ongoing development program, but could be added in the near future.

Nonetheless, the focal point for the next decade will be environmental policy and regulation. At the core is the basic steel producing furnace technology.

GHGs are the Big Story that will lead the development of steel technology over the next decade. The EU is in the lead. There are two choices. Either you can adapt the Blast Furnace, which is further along the road right now. Or, replace the Blast Furnace but this is a longer story.

Sequestration of CO_2, putting it under ground is a major American focus. The US Energy Department supports it. But putting it in the ground may only be a partial and temporary solution.

A paradigm shift in technology will look in a different direction. But, the record for new iron making processes is not good. There are three candidates in the European initiative. They don't reduce CO_2 very much.

Steel Consultant

Electric Arc Furnaces (EAFs) have an inherent advantage among steel-producing facilities because they have a smaller carbon footprint. They use about 30% less carbon to produce a ton of steel. However, the story is more complicated. There is concern that they may just shift the burden to the electricity provider.

International experts in the industry do not see a fundamental breakthrough in steel's carbon footprint any time soon.

We don't see a breakthrough near term. There may be improvements in energy efficiency or synergies between companies that improve net CO_2 results. The combination of better raw materials with new technology can go a long way on better CO_2 results.

Otherwise you are smelting bad stuff.

Steel Consultant

There may be improvements in energy efficiency or synergies between companies that improve net CO_2 results, but this is at existing facilities. More can be done in iron-making but it needs a new site. Co-generation as done in some mines currently could also marginally contribute. More could be done at greenfields but there is no movement likely in this direction from the companies or the public in the near term.

Steel and Recycling

How we view environmental regulation in the steel industry is also affected by how we look at the industry itself. Throughout the book we have looked at the steel industry in terms of supply chains. Production can be thought of not as just a value creation but also as a process of materials transformation in which environmental change and the organization/disorganization of matter and energy are integral rather than incidental to economic activity. In future, it is suggested that we will be seeing supply chains as systems of material flows and balances with material inputs and final outputs in the form of waste and pollution as all part of one system.

By continually reducing energy usage by up to 1% a year, the steel industry has not only met and surpassed the Kyoto Green House Gas targets, it has also made enormous strides in reducing particulates and effluent discharges in the past 20 years. All steel mills, for instance, try to minimize discharges and recycle their water. Steel making uses a lot of water. Some mills have achieved zero discharge; they recycle every drop of water.

The strongest stories come from the EAF mills. Beyond the general claim for steel, which is true in comparison to other industrial materials like aluminum and plastic, there is the record of steel producers' operations themselves and how they have changed in recent years.

> We were the first steel company to have all operations registered to ISO 14001 environmental standards. It shows up in performance. There were always concerns about cooling water and effluent from steel mills. We have no discharge from our mill. We weren't on a waterway so we had to figure out how to minimize water usage. We now have a zero effluent water system. Not even from the washrooms.

Steel Executive

The achievements in steel have literally been remarkable from one end of the ecological story to the other. More importantly, the steel companies have not only cleaned up their own operations but contributed to the clean up and environmental standards for the society and economy as a whole through their scrap retrieval and recycling operations. However much remains to be done, some of it controversial. The record of the steel industry is undeniable. But the conversation is not over.

The steel scrap story requires further elaboration, both not only because EAF steel producers account for approximately half of North American steel production but also because the steel scrap story is an important economic narrative in its own right. It also gives another perspective on how steel contributes to the economy in new and different ways.

The EAFs have a strong green story to tell. They use a lot less energy than BOF mills (25%) and generate a lot less GHGs (10%). Some 98% of their material is recycled.

Steel mills, particularly EAF producers, often have their own scrap divisions or subsidiary companies. The mills use low-grade feedstock for commodity products like rebar. They use recycled auto and appliance material for higher grade products.

High-grade product is available as waste from auto and other manufacturing plants. Shredding comes 60–70% from cars being recycled. Appliances are next and the Loose Material (LOC) is the remainder.

Scrap operations do a value chain analysis of the scrap supply chain. Some material also comes from old buildings and this recovered steel can be endlessly recycled into construction applications. Recycled steel from cars are more limited.

> *We shred 20,000 cars per month. One per minute. The cars are crushed.*
>
> *We take about 400 tons per month from the municipalities, drawing from 70 municipalities.*
>
> *Railcars are another source. We have recently contracted to source 3000 rail cars from a financial services company.*
>
> *Steel Executive*

They also work with dealers and pull product from municipal dump sites. About 10–15% of the feed comes from municipal dump sites. The latter would be made much easier if preliminary sorting was done by consumers through Blue Box sorting in municipalities where households sort their trash into metals, glass, and so on, when they put their garbage out for garbage collection each week.

> *In fact some municipalities are starting to re-mine their dump sites to extract metallics. In municipal dumps, the steel is easily separated because it is magnetic and can be drawn out. Everything else must be hand sorted. Small motors for instance are 88% recyclable. The rest is copper.*
>
> Steel Executive

Contamination issues are critical to the inputs. From the standpoint of the industry, if there were no local steel mills, the material would still have value and be transferred somewhere in the world.

The EU has the most complete recycling program and rules. The life cycle perspective on industrial products, their consumption, and waste management should be a guide for policy for the future across the industrial materials sector.

CHAPTER 11

Challenges and Opportunities

If one accepts that we are moving into a knowledge-based economy then the challenges and opportunities for the steel industry turns on how it learns. The following chapter discusses this in three ways. First, how steel companies learn internally; in particular, how learning takes place on the shop floor by leveraging and enhancing the skills of the steel workforce. Second, given that steel is an intermediate product whose users now function in extended supply chains, what are the learning capabilities that take place in supply chains? Finally, how does steel position itself in the material competition of the future, especially in the major growth market for the future which is construction?

As discussed previously, there are now truly global steel corporations for the first time. What is the new corporate model? Steel and manufacturing are inseparable so their fate largely depends on whether we think that Manufacturing Matters. However, manufacturing itself is now dominated by global supply chains. Within the new collaborative, contingent supply chains, a company's position and success ultimately turns on its learning capabilities.

What are the challenges and opportunities for steel companies going forward?

As the global steel industry finds its place in the new knowledge-based economy, the key variable will be who learns the best and the fastest?

Steel companies are already operating in the new, knowledge-based economy. In the past, being the lowest-cost producer was the key to success. For the future, cost competitiveness will be matched by steel executives' need to build and operate steel companies that are learning organizations.

The problem is not that traditional steel companies don't learn, but they learn only certain things and tend to learn only in one way. It is closely related to how they are organized. Companies also need to learn from their customer bases in new way. Beyond the technical logistics of supply chains, there is the imperative to actively manage the learning capabilities of the firm.

New Steel Company Knowledge Management Strategies

As discussed previously, innovation in steel production technologies has always been. Current steel executives say that for the next decade technological change will be incremental, rather than transformative. No step-function increases in productivity are anticipated. For these reasons, incremental improvement strategies on the shop floor will preoccupy management.

The demographics of the steel industry workforce are heavily skewed to the right end of the age distribution. Having virtually no new hiring for an extended period of time has now created an urgent need to renew the work force.

Share of Employees (All Occupations) over Age 50
in the Broader Steel Sector
1987–2010
Statistics Canada, Labour Force Survey (Special Tabulation)

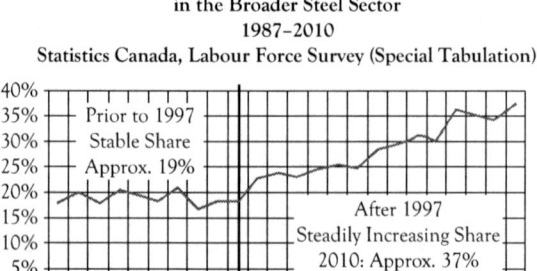

Source: O'Grady & Warrian (2011).

The immediate challenge which keeps steel HR managers up at night is what they identify as the Knowledge Transfer problem.[1]

The skills and knowledge of shop floor workers is rapidly exiting, and production will be compromised as a result. In an actual case, a large steel mill, which is presently managing the exit of 150 employees per year, will come to a standstill if the number jumps to 300. Currently, three times that number are entitled to leave under the pension plan provisions. All steel companies are therefore engaged in major attempts to capture, transfer, and manage the tacit knowledge of production their blue collar work forces have accumulated over the years.

How they do this varies significantly by company and product market objectives. As described below, among the competing approaches is one that seeks to engage production workers at the microeconomic level to reinvigorate continuous improvement. The other is an attempt to capture, codify, and apply workers knowledge in a single standardized global industrial system.

Workers who are over 50—especially lead operators and skilled trades—have acquired knowledge about the production process and the maintenance of machinery and equipment, which is typically undocumented. Transferring this tacit knowledge smoothly and efficiently to new hires is important for maintaining high levels of effective utilization of machinery and equipment. Indeed, for managers, failure to transfer tacit skills and knowledge efficiently could jeopardize effective productivity levels. Given that access to capital and new technology are both dependent on meeting international benchmarking standards, it is strategically important that local steel managers succeed in meeting the knowledge transfer challenge. It is noteworthy that one senior executive in the industry reported that 10% of the company's work force was now engaged in documenting work processes for knowledge transfer purposes.

On the shop floor, two trends will alter the skill requirements of production workers. The first of these is the application of "continuous improvement" (kaizen) to the steel industry. Continuous improvement, as it was pioneered in the Japanese auto industry, is the opposite of traditional steel top-down, engineering-based productivity strategies. In fact, one of the leading steel companies is planning to withdraw all of its engineering staff from the primary steel departments and replace them with community college graduates and production work teams assisted by a new visual management system. This is as big an HR bet as an integrated steelmaker can place.

Continuous improvement in steel seeks cumulative gains in energy use, productivity, and quality from incremental improvements that originate as shop floor initiatives or shop-floor experiments. Studies of companies that have endeavored to implement a "continuous improvement" strategy stress the change in attitude that is required, on the part of both managers and shop-floor workers, for implementation to be successful. Closely linked to the implementation of "continuous improvement" strategies is a flattening of job hierarchies and an expansion of scope within jobs. Narrowly defined jobs do not align well with a "continuous improvement" strategy. All of these developments are challenging the architecture of the CWS system.

The second trend that will incrementally change the skill requirements of production workers is the increased application of information technologies to every aspect of the production process. At the "finishing" end of the steelmaking process, automation will reduce labor requirements and also may reduce reliance on some of the tacit skills that have been accumulated by production workers. At the "front" end of the steelmaking process, automation will increase skill requirements. Over the course of the next 10 years, increased reliance on sensors and computer control systems will make basic computer literacy an essential skill for the majority of steel production workers.

The following are two case studies that illustrate the different approaches steel companies are using to these problems. Eurofer is a large European steel producer. Metallos is an international producer of energy industry products.

Eurofer Case: Kaizen in Steel

Toyota is credited with introducing the "kaizen" or "continuous improvement" strategy into the global manufacturing industry. The process seeks ongoing, incremental improvements in productivity through plant floor problem solving. The "kaizen" approach to productivity improvement reflects its auto industry origins where production processes are initially overengineered and then progressively modified to maximize efficiency by reducing cycle times. There is no simple port over of this approach to a steel plant where engineering decisions have long-term consequences. Nevertheless, the philosophy of "kaizen"—which achieves results through the cumulative impact of small changes—has become central to management strategy in the primary steel industry. The adoption of a continuous improvement strategy is also consistent with the increased emphasis on plant-level, process optimization as the primary focus of R&D efforts and with the application of international benchmarking norms. The adoption of a continuous improvement strategy to achieve long-term, incremental improvements in efficiency has broad implications for work organization and supervisory styles.

All aspects of production, distribution, and management in manufacturing industries are being affected by the adoption of information and communications technologies inputs into the manufacturing sector as a whole. Over the past decade, the merger of information and communications technology now usually called ICT increasingly dominated capital inputs. These trends suggest that the skill profile of the overwhelming majority of occupations will be affected by investments in ICT technologies. This has implications for both occupational skills as well as underscoring the importance of essential skills, for example, reading communications, numeracy, computer literacy, that are a prerequisite to ICT skills.

Eurofer is adopting "Total Productive Maintenance" (TPM) as their principal strategy to achieve and maintain optimal utilization of machinery and equipment. In TPM, a machine operator or team carries out many, and sometimes all, of the routine maintenance tasks. TPM increases the autonomy of equipment operators and operation teams. The objective is to reduce equipment down-time by carrying out regular preventive maintenance. TPM is sometimes linked with the "5-S" philosophy of good maintenance: sort, set in order, systematic cleaning, standardize, and sustain the discipline. The premise of TPM is that machine operators develop tacit knowledge through active management of the machinery with which they work. TPM has important implications for upgrading the skill requirements and responsibilities of equipment operators and for introducing flexible work structures.

In the primary production departments, Eurofer is implementing a visual management. The whole workplace will be set up with signs, labels, color-coded markings, monitors, and so on. Through these visual indicators, anyone unfamiliar with the process can, in a matter of minutes, know what is going on and whether it is being done properly. Visual management is linked to both TPM and continuous improvement.

> Automation may change the skills set at the base level in the front end of the mill. At the finishing end of the mill, automation is generally equivalent to other types of manufacturing, like autos. The most important shift will be into a "visual management system" on the shop floor level.
>
> We intend to drive process improvement from the shop floor. It is a problem solving methodology from Kaizen and leading approaches to training. It requires a mindset change. But it will not be the same in steel as in auto.
>
> Eurofer HR Manager

Clearly this is a major change to the traditional culture of the shop floor in steel. Most interesting will be how the system unfolds in comparison to other examples such as automotive.

Metallos Case: Structured Learning and a Corporate University

The first corporate university was the General Motors Institute (GMI), founded in 1919 as the School of Automotive Trades. GMI provided certification in automotive engineering and also offered degrees in applied sciences and in management tailored to the automotive industry. In 1982, GMI became a private and independent college. In 1997 GMI changed its name to Kettering University. Over the past 20 years, there is evidence of a significant acceleration of the corporate university phenomenon. In 1988, there were 400 corporate training institutions in North America. By 2007, on a worldwide basis, there were approximately 4,000 corporate universities. Corporate universities, also termed academies, institutes, learning centres, or colleges, are established by companies to design and deliver advanced training to their employees and sometimes to their suppliers or users of their products. The substantial investment that goes into these institutions ensures that management is continually aligning training to both the current and long-term human capital needs of the company.

In the steel industry, the leader in establishing and operating a corporate university is Metallos. Many corporate universities restrict their training to middle and upper management, as for instance does Eurofer. Metallos includes the trades and production workers. It's reach is significantly broader, including a wide range of technical and administrative training. Metallos University also plays a key role in the company's overall knowledge management strategy. Ironically, in some respects, this new system is similar in its unitary assumptions with the foundations of the CWS system in the 1940s.

Metallos University consists of six schools—industrial, commercial, finance and administration, management, IT, and technical. All training, including training that is delivered to new employees, is designed by Metallos University. It is unique in that only a small minority of corporate universities offers trades, operator, and technical training. For example, at Metallos University, there are currently more than 30 training plans for mill operators.

Development Plans for Operators and Trades at Metallos (Examples of Job Families)

Operations	Maintenance	Operations support
Steel Making	Maintenance vehicles workshop	Scrap
Heat Treatment	Maintenance—Mechanic/Hydraulic	Equipment
Finishing	Maintenance electric workshop	Logistics and Warehouse
Forming and Welding	Maintenance electronic labs	
Cold Drawn	Maintenance mechanic assembly	
Automotive	workshop	

Training plans are assigned by job families. Each subjob family has five development levels, which represent different levels of skill depth and breadth. After completing an apprenticeship, a tradesperson typically commences at level four. Unskilled workers start at level one.

In addition to its courses, Metallos also delivers highly structured on-the-job training. It is seen as the cornerstone of the intergenerational knowledge transfer strategy. It is purposefully structured to transfer knowledge through practical experience. The strategy also includes training experienced workers to be tutors.

A new employee will go through three stages of on-the-job training:

Job-shadowing an experienced worker. The time spent shadowing varies, depending on the complexity of the job, safety considerations, and risks to quality.

Performing the job under full supervision, 100% of the time. Supervision is given either by a trainer or by a knowledgeable mentor assigned to the new worker.

Performing the job with part-time supervision. The trainees are identified by a band or sticker so everyone on the floor knows who is in training.

As a model for its trades, operator, and technical training, Metallos University looked to Caterpillar University, which has a long history of designing and delivering training to its hourly employees. Caterpillar University remains an important reference point.

Training plans for hourly workers and curricula for salaried employees are developed by Metallos University trainers and experts, using validated best practices as a starting point. Best practices are then distilled into courses and training plans. Further validation occurs through feedback on courses. On an annual basis, Metallos University brings together the supervisors of workers who attended courses. These supervisors are asked what changes were observed in the workplace (whether behavioral, technical, or involving efficiency gains or better awareness around safety). These focus group results serve to further modify and improve courses, which are then provided in the following year.

Metallos University fosters the development of "expertise communities." The knowledge management literature refers to these as "communities of practice." The purpose of these communities of practice is to refine knowledge and accelerate its transfer. Metallos' expertise communities are an adaptation of similar networks introduced by Caterpillar University, which is a benchmark comparator. Also, IBM uses communities of practice as part of its knowledge management strategy. Both Caterpillar and IBM open their communities of practice to suppliers but not to their production employees. Metallos is seeking to expand its "expertise communities" in three areas: maintenance, health and safety, and the environment.

Steel and the Customer Base

Knowledge transfer within and between economic sectors has been identified by economic geographers of innovation as being the key to why some industrial sectors and regions perform above average in revenues, employment and growth over time.

The primary sources of innovation in the steel economy are application of preexisting primary knowledge or application development of technology derived from international networks, that is, global pipelines. This is a classic case of synthetic knowledge transfer.

For steel companies, this means interaction with their manufacturing customers who themselves have undergone qualitative changes through insertion in global supply chains.

In a modern economy, steel and manufacturing have been merged. It is not possible to talk about one without the other. Further, manufacturing has been transformed by global supply chains. The changing nature and dynamics of manufacturing are critical to understanding how knowledge networks operate in the steel industry. They are prime movers in shifting the boundaries of the local industrial economy and a major source of interaction between local and nonlocal actors.

Why study knowledge networks in manufacturing firms? The answer is that they are vital to the future of manufacturing. It is now widely acknowledge that manufacturing, and manufacturing industries across developed countries are now intimately immersed in global supply chains. As the latest scholarship in the area[2] suggests, a firm's position and functioning within global supply chains turns on its learning capabilities. In this sense, the manufacturing economy and the knowledge economy are one.

Manufacturing supply chains and specifically OEM–Supplier relationships have undergone a constant evolution. The classic industrial supply chain was the arm's length market relationship, usually associated with General Motors in its heyday. The job of the supplier firm was simply to meet contract requirements for specified parts and components. The Japanese auto manufacturing revolution brought Toyotism, an integrated, though captive, supplier relationship where there is joint development of design, manufacturing, and quality functions shared between OEM and supplier. This model now dominates across the auto industry and is being copied by other major manufacturing OEMs.

Contract manufacturing takes place at the other end of the spectrum, where design and manufacturing are completely split. The supplier takes on the manufacturing and assembly functions as agreed with the OEM. There are examples in North America, but the classic cases are in Asia, in particular, Taiwan.

Relational contracting is a more integrated form of relationship in which OEM and Supplier form long-term relationships and plan future developments. However, it runs the risk of exclusivity and possible loss of sole sourced contracts.

Industrial studies academics identify Sustained Contingent Collaborative (SCC) as the norm for manufacturing supply chains going forward. In it, OEM and Supplier have close and mutually dependent relations around design, quality, and manufacturing. Both labor under the unrelenting pressure of continuous cost reduction pressures and the need to constantly innovate.

For these reasons, SCC is fundamentally a learning-based process.

A key finding of the research on knowledge networks in auto, steel, and advanced manufacturing was that the innovation that occurs within a sector is greatly determined by how that industry is structured. Factors such as the nature of supplier relationships, the ownership pattern of the industry, and the nature of the product itself play a large role in determining how new products and services are developed and in how R&D resources are allocated. It is striking how differently the innovations within different industries are created and then diffused.

The automotive sector's innovation is entirely dictated by a firm's place in the supply chain. The OEM firms come up with the overall design of a new model and create specifications that they require their part suppliers to fill. However, significant design does take place at the Tier 1 supplier level. The large Tier 1 suppliers have the resources to engage in forward thinking innovations that will allow them to anticipate the OEM's future directions and push technology forward. Innovation also is involved in the automotive sector through the materials providers. While plastic manufacturers and steel mills are not directly involved in the tiering system, the creation of new plastics and steels are integral to making lighter and more efficient cars.

Within the steel sector the consolidation of the industry has resulted in great changes in knowledge flows. Local steel mills have integrated vertically and access information as available within their parent companies' global network. As previously discussed, most "product" innovation in steel takes place in the basic metallurgy of its internal processes. Process and product innovation incubate together in the melt shop. But, steel customers from automotive to construction because they tend to view steel as a commodity, the steel customer base limits, to some extent, the sector's ability to develop new products.

CHAPTER 12

Conclusion

In conclusion, a new global steel industry is emerging, though still it is in its emergent stage. Most of the players and the technology that will be critical in the next decade are already on the field of play. However, there will be plenty of changes within these bounds.

The industry currently stands at a very difficult point in its development. The crisis in Europe, slowdown in China, and less than robust growth in India are all working to create an imbalance in supply and demand that that has taken the steel industry by surprise. Flat rolled prices in North America have fallen from a peak of $740 per ton to about $630 and could bottom out as low as $550.

The European financial crisis has hit steel hard. It is facing a collapse in steel demand with expected demand of 160 MTs but capacity of 220 MTs. Temporary and permanent closures of steel facilities are mounting and assets are being offered for sale at fire sale prices. Thyssen Krupp SA's leading-edge slab processing venture in Brazil and Alabama is said to be in play because of financial problems in its European parent as well a rebounding of competitiveness of domestic US producers who have their own iron ore resources.

The largest steel company in the world, Arcelor, has seen its stock drop 50% in a year. They are idling plants, focusing on their most efficient plants and continued vertical integration in iron ore and coal. There has even been speculation that they may be taken over by a Chinese buyer in the next decade.

The fragile US recovery in steel turns largely on energy development of shale gas and the slow recovery in automotive. Tubular products for the energy industry are booming. Surprisingly, declining gas prices are having an important cost impact on the industry. It is seriously reducing costs for steelmakers. For instance, U.S. Steel consumed more than 100 million BTUs of gas in 2011. This allows them to boost injections of gas into the blast furnace and decrease consumption of coke. The latter should allow steel companies to reduce or even eliminate exposure to the volatile seaborne coal, coke, and iron ore markets. It will also allow consideration of alternative iron and steelmaking capacities such as using iron ore reserves to make directly reduced iron (DRI). Natural gas presents steelmakers with the opportunity to readjust the dynamics of the value chain, which has been tilted toward raw materials for some time.

Within the new internal competition among local managers for capital, there is an even greater emphasis on trying to frame public policies that best support future investment decision-making in the industry. Ideally, there should be a natural, supportive alliance between local steel executives and domestic policy makers.

In addition, there are significant lessons that observers and managers can take up from the steel case and apply to manufacturing and to the economy as a whole.

Steel Industry Restructuring

Further consolidation in steel is inevitable. Otherwise the margins of steel companies will not be sufficient to offset the lack of pricing power upstream to raw material producers and downstream with the automotive industry. In the next decade, we will probably see half dozen steel conglomerates of 100 million tons capacity and all operating in the key markets of North America, Europe, Asia, and the BRIC. The surprises will probably come from the number of emergent Indian and Brazilian companies that will be among that top tier group.

A reverticalization of steel production will take place. Ownership or joint ventures in iron ore, coal, and scrap is to be expected. Closer integration with customers through reacquiring or displacement of service centers will happen, largely because of the need for greater information and technical interchange with final users in advanced manufacturing.

The steel company of the future may look more like Boeing—a systems integrator.

Meanwhile, some service centers and fabricators will evolve to become value-added business service companies.

Materials Competition in the New Economy

A complicated but equally important issue is how steel positions itself for materials competition in the new economy. Certainly in Europe and North America a major growth market will be in the construction sector. In tons of steel consumed, it will probably account in volume for as much as the auto industry. However, materials competition is different in construction than in automotive. It is not clear that this has properly registered on steel management. Materials producers will have to compete not just on metallurgical properties of the product but on technical provisions in Building Codes and the knowledge and skill of construction workers, architects, and civil engineers. At the root, it will be a transition in the business model from physical commodity production to enhanced business services. Steel will essentially become a service business of a specialized kind.

As steel evolves into the materials sector of the new economy, it will be reaching out into related domains. Much greater use of the informational content of steel, that is, "application development" will change the technical competencies and skill requirement at all levels in the steel operations from the shop floor to metallurgical engineers. Steel will have to share the materials space with other materials including carbon, advanced alloys, and other metallic and even steel–plastic composites.

Public Policy and a New Steel Story

When you currently ask North American steel company executives about public policy issues, most list a conventional program of reduced taxes, constrained electricity rates, effective actions on dumped imports and concerns about the pace of environmental regulation. By contrast, the Europeans have a very elaborate European Steel Technology Platform that they are developing and have every intention to implement, to guarantee a place for the industry in the future of their economy.

There is now much commentary in the business media about "on-shoring" and the reindustrialization of America through a renaissance in manufacturing. Because steel is the backbone material, it has to find a place in this narrative. The steel industry will require something more like the European policy example in order to survive and grow in the new economy.

It may sound trite to some engineers but the steel industry has to find some poetry to inject into its story. An example can be seen in the new ArcelorMittal steel exhibit in the Guggenheim Art Museum in Bilbao, Spain.

Lewis Mumford described iron as the material substrate of the emergent urban industrial lifestyle of the 19th century. The most obvious transformation was in the new revolution in transportation of rails, so famously captured in JMW Turner's painting *Steam, Rail and Speed* portraying a locomotive speeding across a bridge through the English country side and capturing the power and thrust of the Industrial Revolution.

The modern lifestyle of the 20th century was built on and of steel. It was steel in the railways, ships bringing the new immigrants to our shores, and steel skyscrapers to our cities. Ironically, the skyscraper emerged as an architectural solution when some brainy engineers turned the design of a steel railway bridge on end to enable them to build taller buildings. They took the idea from Chicago to Manhattan and the rest is history.

We are now all actively engaged in the early stages of debating the directions and implications of a new postindustrial economy and society; this time at a global level.

Some of the components are already clear:

- Light, fuel-efficient cars
- Recycling
- New and renewed energy sources
- Urban design
- New and renewed infrastructure.

For all of these, steel is a critical component.

Steel should be a central part of the materials infrastructure of our future sustainable economy and society. The design and production of the materials we need is only limited by our imagination and dialogue about the environment, lifestyle, and economy we want for ourselves and our children. The materials will be there to match the vision.

In the OECD's scenario for steel, the main issues for public policy will include:

Environmental legislation: achieving effective environmental legislation based on consensus among all players and creating a level playing field so companies can base their decisions on appropriate economic factors.

Labor market policy: elaboration of labor market policies ensuring appropriate support for steel workers, particularly policies for those who might lose their jobs.

Competition policy: to prevent mergers from restricting competition in critical steel product segments as well as access to steel raw materials.

Preservation of markets: to enhance and strengthen existing trade rules to avoid market restrictions and trade frictions resulting from policies in other domains such as environmental regulations.

The OECD Report concludes by pointing out that the future importance of steel industries in all countries will be determined by its' and others' capacity to engage in an effective and inclusive dialogue with other parties in society. The OECD perspective and list of issues provide a useful script for the start of those discussions.

The impact on the economy of the future by the steel industry will not be determined by econometric input/output tables. The steel industry will have a future to the extent that it is able to become a continual, active partner and participant with other social and economic groups in the dialogue about what kind of economy and society we all want for the future.

It is beyond the scope of this book to develop a comprehensive policy framework for steel, let alone the increasing part of that agenda reaching across jurisdictions. That said, the research points to several areas where supportive public policy will advance the competitive conditions for an economically and environmentally vibrant steel industry.

First is to recognize that the "new economy" includes steel. As discussed, steel can indeed be considered an early mover in the knowledge-based economy. The industry is not a 20th-century relic. It is in fact essential to many of the innovations that will drive the economy in the future, for example, conventional and new sources of energy, more fuel-efficient automobiles, enhanced environmental performance, and more efficient life-cycle construction.

The second point is about the competitive environment. Most obviously, this is about competing in domestic and export markets, and ensuring a fair basis for domestic producers to do so, through trade rules and their enforcement.

More generally, it is critical to recognize that the steel industry is now truly global, and investment decisions are also now competitive. In this respect, the industry transformation that resulted in the steel industry becoming part of global multinational enterprises means that local mills must compete for investment capital in that context. From a public policy viewpoint, this means that national governments must offer supportive policies to attract future investment and reinvestment.

This is not a complete or a detailed policy "Agenda" for the steel industry. It is presented here simply to put forth some public policy implications that derive from this research, by indicating key dimensions and types of public policies that can support—or if not addressed, undermine—a sustainable steel industry and its customers. A sound, balanced mix of policy will strengthen the competitive conditions for steelmaking, so that we will continue to benefit from steel's potential as an innovative, competitive industry in the 21st-century economy.

Notes

Chapter 1

1. Helper, Krueger, and Wail (2012).
2. Considine (2012).

Chapter 3

1. Barnett and Schorsch.
2. Chandler (1962).
3. Chandler (1962).
4. Chandler (1990).
5. Prechel (1991).
6. Warren (2001).
7. Warrian (2010); O'Grady and Warrian (2011).
8. Sturgeon (2002, 2008).
9. Sturgeon (2009).
10. Muellerleile (2010).

Chapter 4

1. Herrigel (2010).
2. Yonekura (1994).
3. Barnett and Crandall (1986).
4. Ahlbrandt, Fruehan, and Giarratini (1996).
5. Warrian (2010).

Chapter 5

1. Stubbes (2006); Ginzburg (2009).
2. Kilbourn (1960).
3. Yonekura (1995, 1997).
4. Vincenti (1990).

Chapter 6

1. Stone (1974).
2. Heron (1988).
3. Stieber (1959), p. 26.
4. Stieber (1959).
5. Rose (1998).
6. Rose (1998).
7. Rose (1998).
8. Stieber (1961); Rose (1998).
9. Mangum and McNabb (1997).
10. Becker (1998).
11. Rubinstein (2003) p. 118.
12. Rubinstein (2003) p. 119.
13. Rubinstein (2003) p. 123.
14. Rubinstein (2003).
15. Osterman (2000); Rubinstein (2003).
16. Rubinstein (2001).
17. Bacon et al. (1996).
18. Clark (1993).
19. Clark (1993).

Chapter 7

1. Eichengreen, Van der Ven (1983); Sykes (1996).
2. *China-Measures Related to the Exportation of Various Raw Materials*, WT/DS394/7.
3. United States International Trade Commission, "Antidumping and Countervailing Duty Orders in Place as of February 19, 2010, by Date of Order" (Washington, D.C.: USITC, 2010), online: <http://info.usitc.gov/oinv/sunset.nsf/0a915ada53e192cd8525661a0073de7d/96daf5a6c0c52909852 56a0a004dee7d/$FILE/orders%20February%2019%202010.xls>, accessed March 29, 2010>.
4. Mexico Ministry of the Economy, International Trade Practices Unit, "Sistema de Información sobre las Prácticas Comerciales Internacionales" (Mexico City: Ministry of the Economy, 2010), online: <http://www.pymes.gob.mx/upci/>, accessed March 29, 2010.
5. Sykes (1996).
6. Cooke (2005, 2007a, 2007b).
7. Tony Barber (2012, February 9).

Chapter 8

1. Gereffi, Humphrey, and Sturgeon (2005).
2. Hogan (1971).
3. Aylen (2001).
4. Wells (2001, 2002, 2004).
5. Howard (2000).
6. Rhys (1998).
7. Treado (2008); Treado and Giarratini (2008).
8. Treado (2008).
9. Wixstead (2008).
10. Holweg and Pil (2004).
11. Dooley, Yan, Mohan, and Gopalakrishnan (2010).

Chapter 9

1. Coe (2010).
2. Warrian and Mulhern (2003, 2005).
3. Knoedler and Mayhew (1994).
4. Warrian (2010).

Chapter 10

1. Warren (2001).
2. Thompson (1954).
3. Thompson (1954).

Chapter 11

1. Livingstone and Sawchuk (2004); Livingston, Smith, and Smith (2011).
2. Herrigel (2010).

References

Ahlbrandt, R.S., Fruehan, R.J. & Giarratini, F. (1996). *The renaissance of American steel*. Pittsburgh, PA: University of Pittsburgh Press.

Aylen, J. (2001, July/August). Where did Generation V strip mills come from? *Steel Times*, 225–234.

Bacon, N., Blyton, P. & Jonathan Morris, J. (1996). Among the ashes: Trade union strategies in the UK and German steel industries, *British Journal of Industrial Relations 34*(1), 25–50.

Barnett, D., & Crandall, R. (1986). *Up from the Ashes*. Washington, DC: Brookings Institution.

Becker, G. (1998, March). A history of advocating labor/management cooperation. *New Steel 14*(120).

Chandler, A. (1962). *Strategy and structure*. Cambridge, MA: MIT Press.

Clark, G. (1993). *Pensions and corporate restructuring in American industry: A crisis of regulation*. Baltimore: Johns Hopkins University Press.

Coe, N. (2010). Geographies of production 1: An evolutionary revolution? *Progress in Human Geography 35*(1), 81–91.

Considine, T. (2012). *Economic impacts of the American steel industry*. Washington, DC: American Iron and Steel Institute.

Cooke, P. (2005, October). Regional asymmetric knowledge capabilities and open innovation. *Research Policy, 34*(8), 1128–1149.

Cooke, P. (2007a). Regional innovation, entrepreneurship and talent systems. *International Journal of Entrepreneurship and Innovation Management 7*(2–5), 117–139.

Cooke, P. (2007b). Regional innovation systems, asymmetric knowledge and the legacies of learning. In R. Rutten (Ed.), *The learning region: Foundations, state of the art, future*. Cheltenham: Edward Elgar.

Dooley, K. Yan, T. Mohan, S. & Gopalakrishnan, M. (2010, January). *Journal of Supply Chain Management 46*(1), 12–21.

Eichengreen, B., & Van der Ven, H. (1983). US Anti-Dumping Policies: The Case of Steel. National Bureau of Economic Research, Working paper No. 1098, Cambridge, MA.

Gereffi, G., Humphrey, J., & Sturgeon, T. (2005). The governance of global value chains. *Review of International Political Economy 12*(1), 78–104.

Ginzburg, V. (Ed.). (2009). *Flat rolled steel processes*. Boca Raton, FL: CRC Press.

Heron, C. (1991). *Working in steel: The early years in Canada*. Toronto: McLelland and Stewart.

Herrigel, G. (2010). *Manufacturing possibilities*. Oxford: Oxford University Press.

Hogan, W. (1971). *Economic history of the iron and steel industry in the United States.* Lexington, MA: Heath.

Holweg, M., & Pil, F. (2004). *The second century.* Cambridge, MA: MIT Press.

Howard, M. (2000). Spaceframes. University of Bath, School of Management, Mimeo.

Kilbourn, W. (1960). *The elements combined.* Toronto, ON: Clarke, Irwin & Co.

Knoedler, J. & Mayhew, A. (1994). The engineers and standardization. *Business and Economic History 23*(1), 141–51.

Livingstone, D., & Sawchuk, P. (2004). *Hidden knowledge: Organized labor in the information age.* Toronto, ON: University of Toronto Press.

Livingstone, D., Smith, S., & Smith, W. (2011). *Manufacturing meltdown: reshaping steel work.* Halifax: Fernwood Publishing.

Mangrum, G., & McNabb, R. (1997). *The rise, fall and replacement of industrywide bargaining in the basic steel industry.* Armonk, NY: M.E. Sharpe.

Prechel, H. (1991). Irrationality and contradiction in organizational change: Transformation in the corporate form of a U.S. Steel Corporation. *Sociological Quarterly 32*(3), 423–445.

Osterman, P. (2000). Work reorganization in an era of restructuring: Trends in diffusion and effects on employee welfare. *Industrial and Labor Relations Review 53*(2), 173–187.

Rhys, G. (1998). Steel and the automotive sector: Future prospects. *Steel Times 226*(9), 326.

Rose, J. (1998, September). The struggle over management rights at U.S. steel, 1946–1960: A reassessment of section 2-B of the collective bargaining contract. *Business History Review 72*(3) 446–477.

Rubinstein, S. (2001). A different kind of union: Balancing co-management and representation. *Industrial Relations: A Journal of Economy and Society. 40*(2), 163–203.

Rubinstein, S. (2003). Partnerships of steel? Forging high involvement work systems in the industry: A view from local unions. In (Ed.) 12 *Advances in Industrial & Labor Relations,* (Volume 12), pp. 115–144. Emerald Group Publishing Limited.

Stieber, J. (1959). *The steel industry wage structure.* Cambridge: Harvard University Press.

Stieber, J. (1961). Work rules and practices in mass production industries. *IRRA Proceedings* 399–412.

Stone, K. (1974). The origins of the job structure in the steel industry. *Review of Radical Political Economists 6*(2), 61–97.

Stubbes, J. (2006). The Minimill Story. *AIST J. Keith Brimacombe Memorial Lecture at AISTech 2006.* Cleveland, Ohio.

Sturgeon, T. (2002). How Do We Define Value Chains and Production Networks. MIT Industrial Productivity Centre, Working Paper 00–010.

Sturgeon, T. (2008). From commodity chains to value chains. In Bair, J. (Ed.). *Frontiers of commodity chain research.* Palo Alto, CA: Stanford University Press.

Susan Helper, Timothy Krueger, and Howard Wial. (2012, February). *Why does manufacturing matter? Which manufacturing matters? A Policy Framework.* Washington: Brookings Institution.

Sykes, A. (1996). The economics of injury in anti-dumping and countervailing duty cases. *International Review of Law and Economics 16*(1), 5–26.

Thompson, G. (1954). Intercompany technical standardization in the early American automobile industry. *Journal of Economic History 14*(1), 1–20.

Treado, C. D. (2008, September). *Sustaining Pittsburgh's Steel Technology Cluster. Center for Industry Studies.* University of Pittsburgh.

Treado, C. D., & Giarratini, F. (2008). Intermediate steel-industry suppliers in the Pittsburgh region: A cluster-based analysis of regional economic resilience. *Economic Development Quarterly 22*(1), 63–75.

Warren, K. (2001). *Big steel: The first century of the United States steel corporation 1901–2001.* Pittsburgh, PA: University of Pittsburgh Press.

Warrian, P. (2010). *The importance of steel manufacturing in Canada,* Toronto, ON: Munk School of Global Affairs, University of Toronto.

Warrian, P., & Mulhern C. (2003). Learning in steel: Agents and deficits. In D. A. Wolfe (Ed.), *Clusters old and new: The transition to a knowledge economy in Canada's regions* (pp. 37–62). Montreal & Kingston: McGill-Queens University Press.

Warrian, P., & Mulhern C. (2005). Knowledge and innovation in the interface between the steel and auto industries: The dofasco case. *Regional Studies 39*(2), 161–170.

Wells, P. (2001). Platforms: Engineering panacea, marketing disaster? *Journal of Materials Processing Technology 115,* 166–170.

Wells, P. (2002, March). Cars and wide strip steel: Welded together by history. Centre for Automotive Industry Research, Cardiff Business School, Mimeo.

Wells, P. (2004). Creating sustainable business models: The case of the automotive industry. *IIMB Management Review 16*(4), 15.

Wixted, B. (2008, June). *Cluster rents: Strategic organizations and/or system resources?* DRUID: Copenhagen, CBS, Denmark.

Vincenti, W. (1990). *What engineers know and how they know it.* Baltimore, MD: Johns Hopkins Press.

Yonekura, S. (1994). *The Japanese iron and steel industry 1850–1990.* New York, NY: St. Martin's Press.

Yonekura, S., (1995). Technological innovation in the steel industry: Recognizing potential in innovations. In Minawi, R., Kim, K.S., Makino, F., & Seo, J.H. (Eds.), *Acquiring, adapting and developing technologies,* New York, NY: St. Martin's Press.

Yonekura, S. (1997). The innovation process in the Japanese steel industry. In Goto, A. & Odagiri, H. (Eds.), *Innovation in Japan.* Oxford: Clarendon Press.

Index

OTHER TITLES IN OUR INDUSTRY PROFILES COLLECTION

Donald N. Stengel, Collection Editor

- *A Profile of the Electric Power Industry: Facing the Challenges of the 21st Century* by Charles Clark
- *A Profile of the Oil and Gas Industry: Resources, Market Forces, Geopolitics, and Technology* by Linda Herkenhoff, due Early 2013
- *A Profile of the Wine Industry: Global, Local, Earth, and Glitz* by Barbara Insel, due Fall 2013
- *A Profile of the United States Toy Industry: Serious Fun* by Christopher Byrne, due Fall 2013

Announcing the Business Expert Press Digital Library

Concise E-books Business Students Need for Classroom and Research

This book can also be purchased in an e-book collection by your library as
- a one-time purchase,
- that is owned forever,
- allows for simultaneous readers,
- has no restrictions on printing, and
- can be downloaded as PDFs from within the library community.

Our digital library collections are a great solution to beat the rising cost of textbooks. e-books can be loaded into their course management systems or onto student's e-book readers.

The **Business Expert Press** digital libraries are very affordable, with no obligation to buy in future years. For more information, please visit **www.businessexpertpress.com/librarians**. To set up a trial in the United States, please contact **Adam Chesler** at *adam.chesler@businessexpertpress .com* for all other regions, contact **Nicole Lee** at *nicole.lee@igroupnet.com*.

CPSIA information can be obtained at www.ICGtesting.com
Printed in the USA
BVOW021033220113

311179BV00001B/1/P